Edward Arber

The Dryden Anthology, 1675-1700 A. D.

Edward Arber

The Dryden Anthology, 1675-1700 A. D.

ISBN/EAN: 9783337424244

Printed in Europe, USA, Canada, Australia, Japan

Cover: Foto ©berggeist007 / pixelio.de

More available books at **www.hansebooks.com**

THE

DRYDEN ANTHOLOGY.

1675–1700 A.D.

BRIT ANTH. VII.

BRITISH ANTHOLOGIES.

THE

DRYDEN

ANTHOLOGY.

1675–1700 A.D.

EDITED BY

PROFESSOR EDWARD ARBER, F.S.A.,

FELLOW OF KING'S COLLEGE, LONDON, ETC.

'A thing of beauty is a joy for ever;
Its loveliness increases.'
KEATS.

LONDON:

HENRY FROWDE,

OXFORD UNIVERSITY PRESS WAREHOUSE, AMEN CORNER, E.C.

NEW YORK: 91 & 93 FIFTH AVENUE.

1899.

OXFORD.

HORACE HART, PRINTER TO THE UNIVERSITY.

CONTENTS.

v

Contents.

THE

DRYDEN

ANTHOLOGY.

1675–1700 A. D.

I FEED a flame within, which so torments me
That it both pains my heart, and yet contents me!
'Tis such a pleasing smart, and I so love it;
That I had rather die, than once remove it!

Yet He, for whom I grieve, shall never know it!
My tongue does not betray, nor my eyes show, it!
Not a sigh, nor a tear, my pain discloses;
But they fall silently, like dew on roses.

Thus, to prevent my Love from being cruel,
My heart 's the sacrifice, as 'tis the fuel:
And while I suffer this, to give him quiet;
My faith rewards my love, though He deny it.

On his eyes will I gaze, and there delight me!
While I conceal my love; no frown can fright me!
To be more happy, I dare not aspire;
Nor can I fall more low, mounting no higher!

WHEREVER I am, and whatever I do,
My PHILLIS is still in my mind!
When, angry, I mean not to PHILLIS to go;
My feet, of themselves, the way find!
Unknown to myself, I am just at her door;
And, when I would rail, I can bring out no more
Than, 'PHILLIS, too fair and unkind!'

When PHILLIS I see, my heart bounds in my breast ·
And the love I would stifle, is shown:
But, asleep, or awake, I am never at rest,
When from my eyes PHILLIS is gone!
Sometimes a sad dream does delude my sad mind:
But, alas! when I wake, and no PHILLIS I find;
How I sigh to myself all alone!

Should a King be my rival in her I adore;
He should offer his treasure in vain!
O, let me alone to be happy and poor;
And give me my PHILLIS again!
Let PHILLIS be mine, and but ever be kind;
I could to a desert with her be confined,
And envy no Monarch his reign!

Alas, I discover too much of my love;
And she too well knows her own power!
She makes me, each day, a new martyrdom prove;
And makes me grow jealous each hour!
But let her, each minute, torment my poor mind;
I had rather love PHILLIS, both false and unkind,
Than ever be freed from her power!

THE SEA FIGHT.

WHO ever saw a noble sight,
   That never viewed a brave Sea Fight?
Hang up your bloody colours in the air!
Up, with your fights; and your nettings prepare!
Your merry mates cheer, with a lusty bold spright!
Now each man, his brindice; and then to the fight!
   'St. George! St. George!' we cry!
   The shouting Turks reply!

O, now it begins; and the Gun-room grows hot!
Ply it, with culverin, and with small shot!
Hark! does it not thunder? No! 'tis the guns' roar!
The neighbouring billows are turned into gore!
   Now, each man must resolve to die;
   For, here, the coward cannot fly!
   Drums and trumpets toll the knell;
   And culverins, the passing bell!

Now, now, they grapple; and now board amain!
Blow up the hatches! They're off all again!
Give them a broadside! The dice run at all!
Down come the mast and yard; and tacklings fall!
She grows giddy now, like blind Fortune's Wheel!
She sinks there! She sinks! She turns up her keel!
   Who ever beheld so noble a sight
   As this so brave, so bloody, Sea Fight!

A PASTORAL DIALOGUE
BETWIXT THYRSIS AND IRIS.

FAIR IRIS and her Swain
    Were in a shady bower;
Where THYRSIS long in vain
    Had sought the Shepherd's hour.
At length, his hand advancing upon her
        snowy breast,

THYRSIS.   He said, 'O, kiss me longer,
    And longer yet, and longer,
        If you will make me blest!'

IRIS.   *An easy yielding Maid,*
    By trusting, is undone!
Our sex is oft betrayed
    By granting love too soon!
If you desire to gain me, your suff'rings to
    Prepare to love me longer,   [*redress,*
    And longer yet, and longer,
        Before you shall possess!

THYRSIS.   The little care you show
    Of all my sorrows past,
Makes death appear too slow;
    And life too long to last.

Fair IRIS! kiss me kindly, in pity of my fate,
And kindly still, and kindly,
Before it be too late!

IRIS. *You fondly court your bliss ;*
And no advances make !
'Tis not for Maids to kiss ;
But 'tis for Men to take !
So you may kiss me kindly, and I will not rebel ;
And kindly still, and kindly :
But kiss me not, and tell !

A RONDEAU.

CHORUS. Thus, at the height we love and live ;
And fear not to be poor !
We give, and give, and give, and give,
Till we can give no more !
But what to-day will take away,
To-morrow will restore !
Thus, at the height we love and live ;
And fear not to be poor !

MERCURY'S SONG.

FAIR IRIS I love, and hourly I die ;
But not for a lip, nor a languishing eye !
She 's fickle and false, and there we agree ;
For I am as false and as fickle as she !
We neither believe, what either can say ;
And, neither believing, we neither betray !

5

'Tis civil, to swear, and say things of course;
We mean not the taking for better, for worse!
When present, we love! when absent, agree,
I think not of IRIS; nor IRIS, of me!
The Legend of Love, no couple can find
So easy to part, or so equally joined!

THE TEARS OF AMYNTA,

FOR THE DEATH OF DAMON.

ON a bank, beside a willow,
Heaven her cov'ring, Earth her pillow,
  Sad AMYNTA sighed alone.
From the cheerless dawn of morning
Till the dews of night returning,
  Singing, thus she made her moan.
      'Hope is banished!
      Joys are vanished!
DAMON, my Beloved, is gone!

'Time! I dare thee to discover
Such a Youth! and such a Lover!
  O, so true, so kind, was he!
DAMON was the pride of Nature!
Charming in his every feature!
  DAMON lived alone for me!
      Melting kisses!
      Murmuring blisses!
Who so lived, and loved, as we!' . . .

UNDER THE PORTRAIT OF
JOHN MILTON.

THREE Poets, in three distant Ages born,
Greece, Italy, and England did adorn.
The first, in loftiness of thought surpassed;
The next, in majesty; in both, the last.
The force of Nature could no further go:
To make a Third, she joined the former Two.

———

No! No! Poor suff'ring heart, no change en-
  deavour!
Choose to sustain the smart, rather than leave her!
My ravished eyes behold such charms about her,
I can die with her; but not live without her!
One tender sigh of hers, to see me languish,
Will more than pay the price of my past anguish!
Beware, O, cruel Fair! how you smile on me!
'Twas a kind look of yours that has undone me!

Love has in store for me, one happy minute;
And She will end my pain, who did begin it!
Then, no day void of bliss, or pleasure leaving,
Ages shall slide away without perceiving!
CUPID shall guard the door, the more to please us;
And keep out TIME and DEATH, when they would
  seize us.
TIME and DEATH shall depart; and say, in flying,
'LOVE has found out a way to live, by dying!'

———
7

John Dryden, P.L.

POETS, your subjects, have their Parts assigned,
T' unbend, and to divert, their Sovereign's mind;
When, tired with following Nature, you think fit
To seek repose in the cool shades of Wit:
And from the sweet retreat, with joy survey
What rests, and what is conquered, of the way.
　　Here, free yourselves from envy, care, and strife,
You view the various turns of Human Life!
Safe in our Scene, through dangerous Courts you go;
And, undebauched, the vice of Cities know!
Your theories are here to practice brought,
As in mechanic operations wrought:
And Man, the Little World, before you set;
As once the sphere of crystal showed the Great.
Blest, sure, are you, above all mortal kind,
If to your fortunes you can suit your mind!
Content to see, and shun, those Ills we show;
And crimes, on theatres alone to know!

　　With joy, we bring what our dead Authors writ;
And beg from you the value of their Wit:　[claim
That SHAKESPEARE'S, FLETCHER'S, and great JONSON'S
May be renewed from those who gave them fame!
　　None of our living Poets dare appear;
For Muses so severe are worshipped here,

That, conscious of their faults, they shun the eye,)
And, as profane, from sacred places fly; }
Rather than see th' offended God, and die!)
  We bring no imperfections but our own!
Such faults as made, are by the makers shown;
And you have been so kind, that we may boast,
The greatest Judges still can pardon most!
  Poets must stoop, when they would please our Pit,
Debased even to the level of their wit;
Disdaining that, which yet they know will take;
Hating, themselves, what, their applause must make!
But when, to praise from you they would aspire,
Though they like eagles mount, your JOVE is higher!
So far your knowledge, all their power transcends,
As what SHOULD BE, beyond what IS extends.

ON THE DEATH OF
JOHN CLAVERHOUSE,
EARL OF DUNDEE.

O, LAST and best of Scots! who didst maintain
Thy country's freedom from a foreign reign.
New people fill the Land, now thou art gone!
New Gods, the Temples; and new Kings, the Throne!
  Scotland and thee did in each other live:
Nor wouldst thou, her; nor could she, thee survive!
Farewell! who, dying, did support the State;
And couldst not fall, but with thy country's fate!

9

SONG TO A FAIR YOUNG LADY

GOING OUT OF THE TOWN, IN THE SPRING.

ASK not the cause, why sullen Spring
  So long delays her flowers to bear?
Why warbling birds forget to sing;
  And Winter storms invert the year?
    CHLORIS is gone! and Fate provides
    To make it Spring, where she resides.

CHLORIS is gone! The cruel Fair!
  She cast not back a pitying eye;
But left her Lover in despair,
  To sigh! to languish! and to die!
    Ah! how can those fair eyes endure
    To give the wounds, they will not cure!

Great God of Love! why hast thou made
  A face, that can all hearts command!
That all religions can invade;
  And change the laws of ev'ry land!
    Where thou hadst placed such Power before;
    Thou shouldst have made her Mercy more!

10

When CHLORIS to the Temple comes,
  Adoring crowds before her fall.
She can restore the dead from tombs;
  And ev'ry life but mine recall!
    I only, am by Love designed,
    To be the victim for mankind!

———

SONG,

SUNG TO A MINUET.

How happy the Lover!
  How easy his chain!
  How pleasing his pain!
How sweet to discover
  He sighs not in vain!
For Love, every creature
Is formed by his nature!
  No joys are above
  The pleasures of Love!

In vain are our graces,
  In vain are your eyes,
  If Love you despise!
When Age furrows faces,
  'Tis time to be wise!
Then, use the short blessing
That flies in possessing!
  No joys are above
  The pleasures of Love!

———

11

ALEXANDER'S FEAST;

OR THE POWER OF MUSIC.

PERFORMED AT STATIONERS' HALL, LONDON;

IN HONOUR OF SAINT CECILIA'S DAY, 1697.

I.

'TWAS at the royal feast, for Persia won
  By PHILIP's warlike son.
  Aloft, in awful state,
  The Godlike Hero sat
  On his imperial throne.
His valiant Peers were placed around,
Their brows with roses and with myrtles bound
  (So should desert in Arms be crowned!).
  The lovely THAIS, by his side,
  Sat like a blooming Eastern bride,
  In flower of Youth and Beauty's pride.
  Happy, happy, happy pair!
    None but the Brave,
    None but the Brave,
  None but the Brave deserves the Fair!

CHORUS.

Happy, happy, happy pair !
None but the Brave,
None but the Brave,
None but the Brave deserves the Fair !

II.

TIMOTHEUS, placed on high
  Amid the tuneful Quire,
With flying fingers touched the Lyre.
  The trembling notes ascend the sky,
  And heavenly joys inspire.
The Song began from JOVE,
Who left his blissful seats above,
(Such is the power of mighty Love !).
A dragon's fiery form belied the God,
Sublime on radiant spires he rode,
When he to fair OLYMPIA *pressed ;*
And while he sought her snowy breast.
Then, round her slender waist he curled ;
And stamped an image of himself, a Sovereign of the World.

The list'ning crowd admire the lofty sound.
'A present deity !' they shout around :
'A present deity !' the vaulted roofs rebound.
     With ravished ears,
     The Monarch hears ;
     Assumes the God,
     Affects to nod,
  And seems to shake the Spheres.

13

CHORUS.

With ravished ears,
The Monarch hears;
Assumes the God,
Affects to nod,
And seems to shake the Spheres.

III.

The praise of BACCHUS then, the sweet Musician sung,
Of BACCHUS ever fair, and ever young.
The jolly God in triumph comes!
Sound the trumpets! beat the drums!
Flushed with a purple grace,
He shows his honest face.
Now, give the hautboys breath! He comes! He comes!
BACCHUS, ever fair and young,
Drinking joys did first ordain!
BACCHUS' blessings are a treasure!
Drinking is the soldier's pleasure!
Rich the treasure!
Sweet the pleasure!
Sweet is pleasure after pain!

CHORUS.

BACCHUS' blessings are a treasure!
Drinking is the soldier's pleasure!
Rich the treasure!
Sweet the pleasure!
Sweet is pleasure after pain!

IV.

Soothed with the sound, the King grew vain,
Fought all his battles o'er again ;   [slain.
And thrice he routed all his foes, and thrice he slew the
The Master saw the madness rise,
His glowing cheeks, his ardent eyes ;
And (while he heaven and earth defied)
Changed his hand, and checked his pride.
      He chose a mournful Muse,
      Soft pity to infuse.
He sang, *DARIUS, great and good,*
   By too severe a fate,
Fallen, fallen, fallen, fallen,
   Fallen from his high estate ;
And welt'ring in his blood.
Deserted, at his utmost need,
By those his former bounty fed ;
On the bare earth exposed he lies,
With not a friend to close his eyes.

With downcast looks, the joyless Victor sat,
   Revolving, in his altered soul,
     The various turns of Chance below :
   And, now and then, a sigh he stole ;
     And tears began to flow.

CHORUS.

Revolving, in his altered soul,
  The various turns of Chance below:
And, now and then, a sigh he stole;
  And tears began to flow. .

V.

The mighty Master smiled to see
That Love was in the next degree.
'Twas but a kindred sound to move;
For Pity melts the mind to Love!
  Softly sweet, in Lydian Measures,
  Soon he soothed his soul to pleasures.
War, he sang, *is toil and trouble!*
Honour but an empty bubble!
  Never ending, still beginning,
Fighting still, and still destroying;
  If the world be worth thy winning,
Think, O, think it worth enjoying!
  Lovely THAIS *sits beside thee;*
  Take the good, the Gods provide thee!

The Many read the skies, with loud applause.
So Love was crowned; but Music wan the cause!

The Prince, unable to conceal his pain,
    Gazed on the Fair,
    Who caused his care ;
And sighed and looked, sighed and looked,
Sighed and looked, and sighed again.
At length, with Love and Wine at once oppressed,
The vanquished Victor sank upon her breast.

CHORUS.

The Prince, unable to conceal his pain,
    Gazed on the Fair,
    Who caused his care ;
And sighed and looked, sighed and looked,
Sighed and looked, and sighed again.
At length, with Love and Wine at once oppressed,
The vanquished Victor sank upon her breast.

VI.

Now strike the Golden Lyre again!
A louder yet; and yet a louder strain!
Break his bands of sleep asunder;
And rouse him, like a rattling peal of thunder!
    Hark! Hark! The horrid sound
    Has raised up his head!
    As awaked from the dead,
    And, amazed, he stares around.

Revenge! Revenge! Timotheus *cries,*
See, the Furies arise!
See the snakes that they rear,
How they hiss in their hair;
And the sparkles, that flash from their eyes!
Behold, a ghastly band,
Each a torch in his hand!
Those are Grecian ghosts, that in battle were slain;
And, unburied, remain
Inglorious on the plain.
Give the vengeance due
To the valiant crew!
Behold, how they toss their torches on high!
How they point to the Persian abodes,
And glitt'ring Temples of their hostile Gods!

The Princes applaud, with a furious joy;
And the King seized a flambeau, with zeal to destroy.
Thais led the way,
To light him to his prey;
And, like another Helen, fired another Troy!

Chorus.

And the King seized a flambeau, with zeal to destroy.
Thais led the way,
To light him to his prey;
And, like another Helen, fired another Troy!

18

VII.

Thus, long ago,
Ere heaving bellows learned to blow,
While Organs yet were mute,
TIMOTHEUS, to his breathing Flute
And sounding Lyre,
Could swell the soul to rage ; or kindle soft desire.

At last, divine CECILIA came,
Inventress of the vocal frame.
The sweet enthusiast, from her sacred store,
Enlarged the former narrow bounds,
And added length to solemn sounds,
With Nature's mother-wit, and Arts unknown before.
Let old TIMOTHEUS yield the prize ;
Or both divide the Crown !
He raised a Mortal to the skies !
She drew an Angel down.

GRAND CHORUS.

At last, divine CECILIA came,
Inventress of the vocal frame.
The sweet enthusiast, from her sacred store,
Enlarged the former narrow bounds,
And added length to solemn sounds,
With Nature's mother-wit, and Arts unknown before.
Let old TIMOTHEUS yield the prize;
Or both divide the Crown !
He raised a Mortal to the skies !
She drew an Angel down !

THE LADY'S SONG.

A QUIRE of bright Beauties, in Spring did appear,
To choose a May Lady to govern the year.   [green ;
All the Nymphs were in white, and the Shepherds in
The garland was given, and PHILLIS was Queen.
But PHILLIS refused it ; and, sighing, did say,
' I'll not wear a garland, while PAN is away!

  ' While PAN and fair SYRINX are fled from our
      shore ;
The Graces are banished, and Love is no more!
The soft God of Pleasure, that warmed our desires,
Has broken his bow ; and extinguished his fires :
And vows, that himself and his mother will mourn
Till PAN and fair SYRINX in triumph return!

  ' Forbear your addresses! and court us no more ;
For we will perform what the deity swore!
But if you dare think of deserving our charms,
Away with your sheephooks ; and take to your Arms!
Then laurels and myrtles your brows shall adorn ;
When PAN, and his son, and fair SYRINX return!'

    WHAT state of life can be so blest
    As Love, that warms a Lover's breast!
    Two souls in one! The same desire
    To grant the bliss ; and to require!

But if, in heaven, a hell we find,
    'Tis all from thee,
    O, Jealousy!
    'Tis all from thee,
    O, Jealousy!
Thou tyrant, tyrant, Jealousy!
Thou tyrant of the mind!

All other ills, though sharp they prove,
Serve to refine and perfect Love!
In absence, or unkind disdain,
Sweet Hope relieves the Lover's pain!
But, ah! no cure but death we find,
    To set us free
    From Jealousy!
    O, Jealousy!
Thou tyrant, tyrant, Jealousy!
Thou tyrant of the mind!

False, in thy glass, all objects are:
Some set too near; and some, too far.
Thou art the fire of endless night!
That fire that burns, and gives no light!
All torments of the damned, we find
    In only thee,
    O, Jealousy!
Thou tyrant, tyrant, Jealousy!
Thou tyrant of the mind!

THE SONG OF VENUS

IN REGARD TO BRITAIN.

FAIREST Isle, all isles excelling!
  Seat of Pleasures and of Loves!
VENUS here will choose her dwelling;
  And forsake her Cyprian groves!

CUPID, from his fav'rite nation,
  Care and envy will remove!
Jealousy, that poisons Passion;
  And Despair, that dies for Love!

Gentle murmurs, sweet complaining,
  Sighs that blow the fire of Love,
Soft repulses, kind disdaining,
  Shall be all the pains you prove!

Every Swain shall pay his duty!
  Grateful every Nymph shall prove!
And as these excel in beauty;
  Those shall be renowned for love!

———

  Go, tell AMYNTA, gentle Swain!
I would not die; nor dare complain!
Thy tuneful voice, with Numbers join!
Thy words will more prevail than mine!

To souls oppressed and dumb with grief,
The Gods ordain this kind relief:
That music should, in sounds convey,
What dying Lovers dare not say!

A sigh, or tear, perhaps, she'll give;
But Love, on Pity cannot live!
Tell her, That hearts, for hearts were made;
And love, with love is only paid!
Tell her, My pains so fast increase
That, soon, they will be past redress!
But, ah!, the wretch, that speechless lies,
Attends but Death to close his eyes!

———

HIGH State and honours to others impart;
     But give me your heart!
That treasure, that treasure alone,
     I beg for my own!
So gentle a love, so fervent a fire,
     My soul does inspire;
That treasure, that treasure alone,
     I beg for my own!

Your love, let me crave!
     Give me in possessing
     So matchless a blessing;
That empire is all I would have!

Love 's my petition;
All my ambition!
If e'er you discover
So faithful a Lover,
So real a flame;
I'll die! I'll die,
And give up the game!

———

FAIR, sweet, and young, receive a prize
Reserved for your victorious eyes!
From crowds, whom at your feet you see,
O, pity, and distinguish, me!
As I, from thousand Beauties more
Distinguish you; and only you adore!

Your face for conquest was designed;
For ev'ry motion charms my mind!
Angels, when you your silence break,
Forget their Hymns, to hear you speak!
But when, at once, they hear and view,
Are loth to mount; and long to stay with you!

No graces can your form improve:
But all are lost, unless you love!
While that sweet Passion you disdain,
Your veil and beauty are in vain!
In pity then, prevent my fate!
For, after dying, all reprieve 's too late!

DAMILCAR'S SONG.

Ah! how sweet it is to love!
  Ah! how gay is young Desire!
And what pleasing pains we prove,
    When we first approach Love's fire!
      Pains of Love be sweeter far
      Than all other pleasures are!

Sighs which are from Lovers blown,
  Do but gently heave the heart!
E'en the tears they shed alone,
    Cure, like trickling balm, their smart!
      Lovers, when they lose their breath,
      Bleed away in easy death!

Love and Time, with reverence use!
  Treat them like a parting friend!
Nor the golden gifts refuse,
    Which in Youth sincere they send!
      For, each year, their price is more;
      And they less simple than before!

Love, like Spring-tides full and high,
  Swells in every youthful vein:
But each tide does less supply,
    Till they quite shrink in again.
      If a flow in Age appear;
      'Tis but rain, and runs not clear!

THE SONG OF TRIUMPH

SUNG BY THE VICTORIOUS BRITONS.

COME, if you dare ; our trumpets sound !
Come, if you dare ; the foes rebound !
We come ! We come ! We come ! We come !
Says the double, double, double beat of the thund'ring
    drum.

Now, they charge on amain !
Now, they rally again !
The Gods, from above, the mad labour behold ;
And pity mankind, that will perish for gold.

The fainting Saxons quit their ground !
Their trumpets languish in the sound !
They fly ! They fly ! They fly ! They fly !
'Victoria ! Victoria !' the bold Britons cry.

'Now the victory 's won ;
To the plunder we run !
We return to our Lasses, like fortunate traders,
Triumphant with spoils of the vanquished invaders !' ·

A RONDELAY.

CHLOE found AMYNTAS lying,
    All in tears, upon the plain,
Sighing to himself, and crying,
    'Wretched I, to love in vain!
Kiss me, Dear! before my dying!
    Kiss me once, and ease my pain!'

Sighing to himself, and crying,
    'Wretched I, to love in vain!
Ever scorning, and denying
    To reward your faithful Swain!
Kiss me, Dear! before my dying!
    Kiss me once, and ease my pain!

'Ever scorning, and denying
    To reward your faithful Swain!'
CHLOE, laughing at his crying,
    Told him, That he loved in vain.
'Kiss me, Dear! before my dying!
    Kiss me once, and ease my pain.'

CHLOE, laughing at his crying,
    Told him, That he loved in vain:
But repenting, and complying;
    When he kissed, she kissed again.
Kissed him up, before his dying;
    Kissed him up, and eased his pain.

THE FAIR STRANGER.

HAPPY and free, securely blest,
No Beauty could disturb my rest!
My amorous heart was in despair
To find a new victorious Fair.

Till you, descending on our plains,
With foreign force renew my chains;
Where now you rule without control,
The mighty Sovereign of my soul.

Your smiles have more of conquering charms
Than all your native country's Arms!
Their troops we can expel with ease;
Who vanquish only when we please.

But in your eyes, O, there 's the spell!
Who can see them, and not rebel!
You make us captives by your stay;
You kill us, if you go away!

———

AII! fading Joy! how quickly art thou past!
    Yet we thy ruin haste!
As if the cares of Human Life were few,
    We seek out new;
And follow Fate, that does too fast pursue!

See, how on every bough the birds express,
    In their sweet notes, their happiness!
    They all enjoy, and nothing spare;
But on their mother, Nature, lay their care.
Why then should Man, the Lord of all below,
    Such troubles choose to know,
As none of all his subjects undergo?

Hark! Hark! The waters fall, fall, fall!
    And, with a murmuring sound,
    Dash, dash upon the ground,
      To gentle slumbers call.

'VENI, CREATOR SPIRITUS!'

TRANSLATED IN PARAPHRASE.

CREATOR SPIRIT! by whose aid
The world's foundations first were laid,
Come, visit ev'ry pious mind!
Come, pour thy joys on human kind!
From Sin and Sorrow set us free;
And make thy Temples worthy Thee!

O, Source of uncreated Light!
The Father's promised Paraclete!
Thrice-holy Fount! thrice-holy Fire!
Our hearts with heavenly love inspire!
Come, and thy sacred unction bring
To sanctify us, while we sing!

Plenteous of Grace, descend from high,
Rich in thy sevenfold energy!
Thou strength of His almighty hand;
Whose power doth Heaven and Earth command.
Proceeding SPIRIT! our defence!
Who doth the Gift of Tongues dispense;
And crown'st thy gift, with eloquence!

Refine and purge our earthy parts;
But O, inflame and fire our hearts!
Our frailties help! our vice control!
Submit the Senses to the Soul!
And when rebellious they are grown;
Then lay thy hand, and hold them down!

Chase from our minds th' Infernal Foe;
And Peace, the fruit of Love, bestow!
And, lest our feet should step astray,
Protect and guide us in the way!

Make us eternal truths receive;
And practise all that we believe!
Give us Thyself! that we may see
The Father and the Son, by Thee!

Immortal honour, endless fame,
Attend th' Almighty Father's name!
The Saviour Son be glorified!
Who for lost Man's redemption died:
And equal adoration be,
Eternal Paraclete, to Thee!

MAXIMINIAN'S SONG.

WHAT shall I do, to show how much I love her?
　How many millions of sighs can suffice?
That which wins other hearts, never can move her!
　Those common methods of Love she'll despise!

I will love more than Man e'er loved before me!
　Gaze on her all the day; melt all the night!
Till, for her own sake, at last, she'll implore me
　To love her less, to preserve our delight!

Since Gods themselves could not ever be loving,
　Men must have breathing recruits for new joys!
I wish my love could be always improving;
　Though eager love, more than sorrow, destroys!

In fair AURELIA's arms, leave me expiring;
　To be embalmed by the sweets of her breath!
To the last moment, I'll still be desiring!
　Never had Hero so glorious a death!

PEACE, CUPID! Take thy bow in hand!
I' th' gloomy shade, in ambush stand,
To watch a cruel Nymph frequents this bower,
Cold as the streams; but sweeter than each flower!
There, there, She is! Direct thy dart
Into that stony marble heart!
Draw, quickly draw; and shew thy art!
Woe 's me! Thou art blind indeed! Thou hast
shot me!
While She 'scapes in the grove; and laughs at thee.

Is CELADON unkind? It cannot be!
Or is he so unconstant grown
To slight my vows; and break his own?
Forbid it, Heaven! No! it cannot be!
Then, my good Angel, whither is he fled?
Tell me! O, tell me softly! Is he dead?
Ah! prophetic soul, forbear!
Lest I languish in despair.
No! My heart, whene'er he dies,
In the pain must sympathize!
Since my soul and his are one,
He cannot live, or die, alone!
FLORELLA! forbear to distrust, or repine!
Since his love, and his suff'rings, are equal with thine.
And when he returns, if ever again,
We'll kiss away Sorrow; and laugh away Pain!

A FAREWELL
TO THE VANITIES OF THE WORLD.

FAREWELL, ye gilded follies! pleasing troubles!
Farewell, ye honoured rags! ye glorious bubbles!
Fame 's but a hollow echo! gold, pure clay!
Honour, the darling but of one short day!
Beauty, th' eyes' idol, but a damasked skin!
State, but a golden prison, to live in,
And torture free-born minds! Embroidered Trains,
Merely but pageants for proud swelling veins!
And Blood allied to Greatness is alone
Inherited; not purchased! nor our own!
    Fame, Honour, Beauty, State, Train, Blood, and Birth,
    Are but the fading blossoms of the Earth!

I would be *great*; but that the sun doth still
Level his rays against the rising hill.
I would be *high*; but see the proudest oak
Most subject to the rending thunder-stroke.
I would be *rich*; but see men (too unkind!)
Dig in the bowels of the richest mind.
I would be *wise*; but that I often see
The fox suspected, whilst the ass goes free.
I would be *fair*; but see the fair and proud,
Like the bright sun, oft setting in a cloud.
I would be *poor*; but know the humble grass
Still trampled on by each unworthy ass.
Rich, hated! wise, suspected! scorned, if poor!
Great, feared! fair, tempted! high, still envied more!

34

I have wished all; but now I wish for neither
Great, high, rich, wise, nor fair; poor, I'll be rather!

Would the World now adopt me for her heir!
Would Beauty's Queen entitle me, the fair!
Fame speak me Fortune's minion! Could I vie
Angels with India! with a speaking eye,   [dumb,
Command bare heads, bowed knees! strike Justice
As well as blind, and lame! or give a tongue
To stones by Epitaphs! be called 'great Master!'
In the loose rhymes of every Poetaster!
    Could I be more than any man that lives,
Great, fair, rich, wise; all in superlatives!
Yet I more freely would these gifts resign,
Than ever Fortune would have made them mine;
And hold one minute of this holy leisure
Beyond the riches of this empty pleasure!

Welcome, pure thoughts! Welcome, ye silent groves!
These guests, these Courts, my soul most dearly loves!
Now the winged people of the sky shall sing
My cheerful Anthems to the gladsome Spring!
A *Prayer Book* now shall be my looking-glass;
In which I will adore sweet Virtue's face!
Here, dwell no hateful looks; nor palace cares!
No broken vows dwell here; nor pale-faced fears!
Then here I'll sit, and sigh my hot love's folly;
And learn t' affect a holy melancholy!
    And if Contentment be a stranger then;
I'll ne'er look for it, but in Heaven again!

Why, Cloris! should you on him frown,
  Who always owned your power?
The glory of that Triumph 's gone,
Where no resistance could be shown!
  I was your slave before!

May some more noble enterprise
  Your charming force engage!
Such as rebel against your eyes,
Or dare your haughty mien despise,
  Are objects for your rage!

So you, your empire may advance;
  And you, secure your reign!
For thus, your conquest you'll enhance;
While some great captive, every glance
  Reduces to your chain!

But while you, 'midst your Trophies are;
  Scorn not your loyal slave!
For should I, equal penance share,
With those that once rebellious were;
  'Twould bring me to my grave!

ERICINIA'S SONG.

O, YOU powerful Gods! If I must be
An injured offering to LOVE's deity;
Grant my revenge! this plague on Men!
That Women ne'er may love again!
Then I'll, with joy, submit unto my fate;
Which, by your justice, gives their empire date.

Depose that proud insulting Boy;
Who most is pleased, when he can most destroy!
O, let the World no longer governed be
By such a blind and childish deity!
For if you Gods be in your power severe,
We shall adore you, not from love, but fear!

But if you'll his divinity maintain;
O'er men, false men! confine his tort'ring reign!
And when their hearts, LOVE's greatest torments prove,
Let that, not pity; but our laughter, move!
Thus scorned, and lost to all their wishes' aim;
Let Rage, Despair, and Death then end their flame

LADIES! farewell! I must retire!
Though I, your faces all admire,
And think you Heavens in your kinds,
Some for beauties, some for minds:
If I stay, and fall in love,
One of these Heavens, Hell would prove.

Could I know one, and she not know it;
Perhaps I then might undergo it!
But if the least She guess my mind;
Straight in a circle I'm confined!
By this I see, who once doth dote,
Must wear a woman's livery coat!

Therefore, this danger to prevent,
And still to keep my heart's content;
Into the Country I'll with speed!
With hounds and hawks my fancy feed
Both safer pleasures to pursue,
Than staying to converse with you.

THE SOLDIERS' SONG.

'To Arms! To Arms!' the Heroes cry,
  'A glorious death, or victory!'
Beauty and Love, although combined,
  And each so powerful alone,
Cannot prevail against a mind
  Bound up in resolution!
Tears, their weak influence vainly prove;
Nothing the daring breast can move!
Honour is blind, and deaf! e'en deaf to Love!

The Field! the Field! where Valour bleeds,
Spurned into dust by barbèd steeds,
Instead of wanton beds of down,
  Is now the scene, where we must try
To overthrow; or be o'erthrown!
  Bravely to overcome, or die!
Honour, in her int'rest, sits above
What Beauty, Prayers, or Tears, can move!
Were there no Honour; there would be no Love!

REASON and TIME had once agreed,
My heart from loving should be freed:
But CUPID swore, he'd lay a snare
Should catch my reason, time repair!
SYLVIA appeared! with all the charms
   And witchcrafts of a face
Able to do all Mankind harms;
   And Womankind disgrace!

REASON straight fled! TIME would have stayed,
   Mistaking for the sun
The glories of the brighter Maid;
   By those, his course to run.

JOVE saw, and feared some strange surprise;
   Lest all the World should be
Immortal made by her bright eyes;
   And scorn his deity.
So TIME was forced to fly; Old Age, remain:
But, ah! poor REASON ne'er came back again!

———

LET us turn usurers of time;
And not misspend an hour!
The present, not the future, 's in our power.
To think to spend what 's not our own, 's a crime!
He whose soft life 's in mirth possest,
Enjoys his time with interest!

Love and a Muse
Bring use on use!
For Money 's but the Slave; and Time, the Measure;
And Wit, the Handmaid; Love, the Queen of Pleasure!

———

CUPID! I scorn to beg the art
From thy imaginary throne,
To learn to wound another's heart;
Or how to heal my own!
If She be coy; my airy mind
Brooks not a siege! If She be kind;
She proves my scorn, that was my wonder!
For towns that yield, I hate to plunder!

Love is a game, Hearts are the prize,
Pride keeps the stakes, Art throws the dice.
When either 's won,
The game is done!
Love is a coward, hunts the flying prey;
But when it once stands still, Love runs away!

THE ADVICE.

PHYLLIS, for shame, let us improve
  A thousand several ways.
These few short minutes, stolen by love
  From many tedious days!

Whilst you want courage to despise
  The censure of the Grave,
For all the tyrants in your eyes,
  Your heart is but a slave!

My love is full of noble pride,
  And never will submit
To let that Fop, Discretion, ride
  In triumph over Wit!

False friends I have, as well as you,
  That daily counsel me,
Vain frivolous trifles to pursue;
  And leave off loving thee!

When I the least belief bestow
  On what such fools advise;
May I be dull enough to grow
  Most miserably wise!

———

Charles Sackville, Earl of Dorset.

MAY the ambitious ever find
  Success in Courts and noise;
While gentle Love does fill my mind
  With silent real joys!

May Knaves and Fools grow rich and great,
  And the World think them wise;
While I lie dying at her feet,
  And all that World despise!

Let conquering Kings. new triumphs raise;
  And melt in Court delights
Her eyes can give much brighter days;
  Her arms, much softer nights!

———

THE fire of Love in youthful blood,
Like what is kindled in brushwood,
  But for a moment burns!
Yet, in that moment. makes a mighty noise!
  It crackles, and to vapour turns;
  And soon itself destroys!

But when crept into agèd veins,
It slowly burns. and long remains,
  And with a sullen heat:
Like fire in logs. it glows and warms them long!
  And though the flame be not so great;
    Yet is the heat as strong!

———

43

DORINDA's sparkling wit and eyes
    United, cast so fierce a light;
Which blazes high, then quickly dies!
    Warms not the heart, but hurts the sight!
True Love, all gentleness and joy,
    Approaches with a modest grace:
Her CUPID is a blackguard boy,
    That claps his link full in your face!

—

THE dainty young heiress of Lincoln's Inn Fields,
    Brisk, beautiful, wealthy, and witty,
To the power of Love so unwillingly yields,
    That 'tis feared she'll unpeople the City!
The Sparks and the Beaus all languish and die:
    Yet, after the conquest of many,
One little good marksman, that aims with one eye,
    May wound her heart deeper than any!

44

THE AUTHOR'S

APOLOGY FOR HIS BOOK.

['*The Pilgrim's Progress,' Part I, 1678.*]

WHEN, at the first, I took my pen in hand
Thus for to write; I did not understand
That I at all should make a little book
In such a mode. Nay! I had undertook
To make another: which, when almost done,
Before I was aware, I this begun.
　And thus it was. I, writing of the Way
And Race of Saints in this our Gospel Day,
Fell suddenly into an allegory
About their Journey, and the Way to Glory,
In more than twenty things; which I set down.
This done; I twenty more had in my crown:
And they again began to multiply
Like sparks that from the coals of fire do fly.
　'Nay, then,' thought I, 'if that you breed so fast,
I'll put you by yourselves! lest you, at last,
Should prove *ad infinitum*; and eat out
The Book that I already am about!'

　Well! so I did. But yet I did not think
To show to all the World my pen and ink
In such a mode. I only thought to make
I knew not what! Nor did I undertake
Thereby to please my neighbour. No! not I!
I did it mine own self to gratify.

Neither did I, but vacant seasons spend
In this my Scribble: nor did I intend
But to divert myself, in doing this,
From worser thoughts; which make me do amiss.
Thus, I set pen to paper with delight;
And quickly had my thoughts in black and white:
For, having now my method by the end,
Still as I pulled, it came! And so I penned
It down, until it came, at last, to be,
For length and breadth, the bigness which you see.

Well! when I had thus put mine ends together,
I showed them others; that I might see, whether
They would condemn them; or them justify.
And some said 'Let them live!' some, 'Let them die!'
Some said 'JOHN, print it!' Others said 'Not so!'
Some said 'It might do good!' Others said 'No!'
Now was I in a strait; and did not see
Which was the best thing to be done by me.
At last, I thought, 'Since you are thus divided,
I print it will!' and so the case decided.
'For,' thought I, 'some, I see, would have it done;
Though others in that channel do not run.'
To prove then, who advisèd for the best,
Thus I thought fit to put it to the test.
I further thought, 'If now I did deny
Those that would have it thus, to gratify;
I did not know, but hinder them I might
Of that which would to them be great delight.'

For those that were not for its coming forth,
I said to them, 'Offend you I am loth:
Yet since your brethren pleasèd with it be;
Forbear to judge, till you do further see!
'If that thou wilt not read; let it alone!
Some love the meat; some love to pick the bone!
Yea, that I might them better palliate,
I did too, with them thus expostulate.
'May I not write in such a style as this,
In such a method too; and yet not miss
Mine end, thy good? Why may it not be done?
Dark clouds bring waters, when the bright bring none!
Yea, dark or bright, if they their silver drops
Cause to descend, the earth, by yielding crops,
Gives praise to both; and carpeth not at either:
But treasures up the fruit they yield together;
Yea, so commixes both, that, in her fruit,
None can distinguish this from that. They suit
Her well, when hungry; but if she be full,
She spews out both, and makes their blessings null.
'You see the ways the Fisherman doth take
To catch the fish. What engines doth he make!
Behold, how he engageth all his wits,
Also his snares, lines, Angles, hooks, and nets;
Yet fish there be, that neither hook, nor line,
Nor snare, nor net, nor engine, can make thine!
They must be groped for, and be tickled too;
Or they will not be catched, whate'er you do!
'How doth the Fowler seek to catch his game
By divers means, all which one cannot name!

His gun, his nets, his lime-twigs, light, and bell!
He creeps! He goes! He stands! Yea, who can tell
Of all his postures! Yet there 's none of these
Will make him master of what fowls he please!
Yea, he must pipe, and whistle, to catch this;
Yet if he does so, that bird will he miss!
  'If that a pearl may in a toad's head dwell,
And may be found too in an oyster shell;
If things that promise nothing do contain
What better is than gold: who will disdain
(That have an inkling of it) there to look,
That they may find it. Now, my little Book
(Though void of all those paintings that may make
It with this, or the other, man to take)
Is not without those things that do excel,
What do in brave, but empty, notions dwell.'

  'Well, yet I am not fully satisfied,
That this your Book will stand; when soundly tried.'

  'Why! what 's the matter?' *'It is dark!'* 'What tho?'
'But it is feigned!' 'What of that, I trow!
Some men, by feigning words as dark as mine,
Make Truth to spangle, and its rays to shine!'

  'But they want solidness!' 'Speak, man, thy mind!'
'They drown the weak! Metaphors make us blind!'
  48

'Solidity, indeed, becomes the pen
Of him that writeth things divine to men:
But must I needs want solidness; because
By metaphors I speak? Were not GOD's Laws,
His Gospel Laws, in older time, held forth
By types, shadows, and metaphors? Yet loth
Will any sober man be to find fault
With them; lest he be found for to assault
The highest Wisdom! No! he rather stoops,
And seeks to find out what, by pins and loops,
By calves and sheep, by heifers and by rams,
By birds and herbs, and by the blood of lambs,
GOD speaketh to him! And happy is he
That finds the light and grace that in them be!
    'Be not too forward therefore to conclude
That I want solidness! that I am rude!
All things solid in show, not solid be!
All things in parables despise not we!
Lest things most hurtful lightly we receive;
And things that good are, of our souls bereave.
    'My dark and cloudy words, they do but hold
The Truth, as cabinets inclose the gold.
    'The Prophets usèd much by metaphors
To set forth Truth! Yea, whoso considers
CHRIST, his Apostles too, shall plainly see
That Truths, to this day, in such mantles be.
    'Am I afraid to say, that Holy Writ,
Which for its style and phrase puts down all Wit,
Is everywhere so full of all these things
(Dark Figures, Allegories): yet there springs

From that same Book, that lustre, and those rays
Of light, that turn our darkest nights to days.

'Come, let my Carper, to his life now look;
And find there, darker lines than in my Book,
He findeth any! Yea, and let him know,
That, in his best things, there are worse lines too!
    'May we but stand before impartial men;
To his poor One, I durst adventure Ten!
That they will take my meaning in these lines
Far better than his lies in silver shrines!
Come, Truth; although in swaddling clouts! I find
Informs the judgement, rectifies the mind,
Pleases the understanding, makes the will
Submit! The memory too, it doth fill
With what doth our imagination please!
Likewise, it tends our troubles to appease.
    '"Sound words," I know TIMOTHY is to use;
And "old wives' fables" he is to refuse;
But yet grave PAUL him nowhere doth forbid
The use of Parables! in which lay hid
That gold, those pearls, and precious stones that were
Worth digging for; and that, with greatest care!

'Let me add one word more, O, Man of GOD!
Art thou offended? Dost thou wish I had
Put forth my matter in another dress;
Or that I had, in things, been more express?

Three things let me propound; then I submit
To those that are my betters, as is fit.
　1. 'I find not that I am denied the use
Of this my method; so I no abuse
Put on the Words, Things, Readers; or be rude
In handling Figure, or Similitude
In application: but, all that I may,
Seek the advance of Truth, this, or that, way.
"Denied," did I say? Nay! I have leave
(Example too; and that from them that have
GOD better pleasèd by their words, or ways,
Than any man that breatheth now-a-days!)
Thus to express my mind; thus to declare
Things unto thee, that excellentest are.
　2. 'I find that men (as high as trees) will write
Dialogue-wise: yet no man doth them slight
For writing so! Indeed, if they abuse
Truth; cursed be they, and the craft they use
To that intent! But yet let Truth be free
To make her sallies upon thee, and me,
Which way it pleases GOD! For who knows how,
Better than He that taught us first to plow,
To guide our mind and pens for his design!
And He makes base things usher in divine!
　3. 'I find that Holy Writ, in many places,
Hath semblance with this method; where the cases
Doth call for one thing to set forth another.

'Use it I may then; and yet nothing smother
Truth's golden beams! Nay! by this method, may
Make it cast forth its rays as light as day!
  'And now, before I do put up my pen,
I'll shew the profit of my Book; and then
Commit both thee, and it, unto that Hand   [stand.
That pulls the strong down; and makes weak ones

  'This Book it chalketh out before thine eyes
The Man that seeks the Everlasting Prize.
It shews you, whence he comes; whither he goes;
What he leaves undone; also what he does.
It also shews you, how he runs, and runs,
Till he unto the Gate of Glory comes.
  'It shews too, who sets out for life amain
As if the lasting crown they would attain.
Here also, you may see the reason why
They lose their labour; and like fools do die.
  'This Book will make a Traveller of thee;
If by its counsel, thou wilt rulèd be!
It will direct thee to the Holy Land;
If thou wilt its directions understand!
Yea, it will make the slothful, active be;
The blind also, delightful things to see!
  'Art thou for something rare and profitable?
Wouldst thou see a Truth within a Fable?
Art thou forgetful? Wouldst thou remember
From New Year's Day to the last of December?
Then read my Fancies! They will stick like burrs;
And may be to the helpless, comforters.

52

'This Book is writ in such a dialect
As may the minds of listless men affect!
It seems a novelty; and yet contains
Nothing but sound and honest Gospel strains.
 'Wouldst thou divert thyself from melancholy?
Wouldst thou be pleasant; yet be far from folly?
Wouldst thou read riddles, and their explanation;
Or else be drownèd in thy contemplation?
Dost thou love picking meat? or wouldst thou see
A man i' th' clouds; and hear him speak to thee?
Wouldst thou be in a Dream, and yet not sleep;
Or wouldst thou, in a moment, laugh and weep?
Wouldst thou lose thyself, and catch no harm;
And find thyself again without a charm?
Wouldst read thyself, and read thou know'st not what;
Aud yet know, whether thou art blest, or not,
By reading the same lines? O, then come hither:
And lay my Book, thy head, and heart, together!'

JOHN BUNYAN.

———

THE trials, that those men do meet withal
That are obedient to the heavenly call,
Are manifold, and suited to the flesh;
And come, and come, and come again afresh,
That now, or some time else, we by them may
Be taken, overcome, and cast away.
O, let the Pilgrims, let the Pilgrims then
Be vigilant; and quit themselves like men!

———

Rev. John Bunyan.

THE SHEPHERD BOY'S SONG,

IN THE VALLEY OF HUMILIATION.

HE that is down, needs fear no fall!
    He that is low, no pride!
He that is humble, ever shall
    Have GOD to be his guide!

I am content with what I have,
    Little be it, or much!
And, LORD! contentment still I crave;
    Because Thou savest such!

Fullness to such, a burden is,
    That go on pilgrimage.
Here, little; and hereafter, bliss,
    Is best from Age to Age!

PRITHEE, why so angry? Sweet!
　　'Tis, in vain,
　To dissemble a disdain!
That frown i' th' infancy I'll meet;
　And kiss it to a smile again!

In that pretty anger is
　　　Such a grace,
　As Love's fancy would embrace,
As to new crimes may Youth entice;
　So that disguise becomes that face!

When thy rosy cheek thus checks
　　　My offence,
　I could sin with a pretence;
Though that sweet chiding blush there breaks
　So fair, so bright an innocence!

Thus, your very frowns entrap
　　　My desire;
　And inflame me to admire
That eyes, dressed in an angry shape,
　Should kindle, as with amorous fire!

———

THE RETIREMENT.

TO MASTER ISAAK WALTON.

FAREWELL, thou busy World! and may
 We never meet again!
Here, I can eat, and sleep, and pray,
And do more good in one short day,
Than he who his whole age outwears
Upon the most conspicuous theatres;
Where nought but vanity and vice appears!

Good God! how sweet are all things here!
How beautiful the fields appear!
How cleanly do we feed and lie!
Lord! what good hours do we keep!
 How quietly we sleep!
What peace! What unanimity!
How innocent from the lewd fashion
Is all our business! all our recreation!

O, how happy here 's our leisure!
O, how innocent, our pleasure!
O, ye valleys! O, ye mountains!
O, ye groves and crystal fountains!
   How I love, at liberty,
By turns, to come and visit ye!

Dear Solitude, the soul's best friend!
That Man, acquainted with himself dost make;
And all his Maker's wonders to intend!
   With thee, I here converse at will;
   And would be glad to do so still!
For it is thou alone, that keep'st the soul awake!

How calm and quiet a delight
     Is it, alone,
To read, and meditate, and write!
By none offended; and offending none!
To walk, ride, sit, or sleep, at one's own ease!
And pleasing a man's self; none other to displease!

O, my belovèd Nymph, fair Dove!
Princess of rivers! how I love
Upon thy flow'ry banks to lie,
   And view thy silver stream,
   When gilded by a summer's beam!
And, in it, all thy wanton fry

Playing at liberty!
And, with my Angle, upon them
The all of treachery
I ever learned, industriously to try!

Such streams, Rome's yellow Tiber cannot show;
The Iberian Tagus, or Ligurian Po!
The Maas, the Danube, and the Rhine,
Are puddle water all, compared with thine!
And Loire's pure streams, yet too polluted are,
With thine, much purer, to compare!
The rapid Garonne and the winding Seine
Are both too mean,
Belovèd Dove! with thee
To vie priority!
Nay, Tame and Isis, when conjoined, submit;
And lay their trophies at thy silver feet!

O, my belovèd rocks! that rise
To awe the earth, and brave the skies!
From some aspiring mountain's crown,
How dearly do I love,
Giddy with pleasure, to look down!
And from the vales, to view the noble heights above!
O, my belovèd caves! from dog-star's heat,
And all anxieties, my safe retreat!
What safety, privacy; what true delight,

In th' artificial night
Your gloomy entrails make,
Have I taken! do I take!
How oft (when grief has made me fly,
To hide me from society,
Even of my dearest friends) have I,
In your recesses' friendly shade,
All my sorrows open laid;
And my most secret woes entrusted to your privacy!

Lord! would men let me alone;
What an over-happy one,
Should I think myself to be!
Might I, in this desert place,
(Which most men in discourse disgrace),
Live but undisturbed and free!
Here, in this despised recess,
Would I (maugre Winter's cold
And Summer's worst excess)
Try to live out, to sixty full years old!
And all the while
(Without an envious eye
On any, thriving under Fortune's smile)
Contented, live! and then, contented, die!

———

THE ANGLER'S BALLAD.

Away to the brook!
All your tackle out look!
Here 's a day that is worth a year's wishing!
See that all things be right!
For 'tis very spite
To want tools, when a man goes a fishing!

Your rod with tops two;
For the same will not do,
If your manner of angling do vary!
And full well you may think,
If you troll with a pink,
One too weak will be apt to miscarry!

Then, basket, neat made
By a master in 's trade,
In a belt at your shoulders must dangle:
For none e'er was so vain,
To wear this to disdain,
Who a true Brother was of the Angle.

Next, pouch must not fail!
Stuffed as full as a mail
With wax, cruels, silks, hair, furs, and feathers,
To make several flies
For the several skies;
That shall kill, in despite of all weathers.

The boxes and books
For your lines and your hooks;
And, though not for strict need notwithstanding,
Your scissors and your hone
To adjust your points on;
With a net, to be sure, for your landing.

All these being on;
'Tis high time we were gone!
Down, and upward, that all may have pleasure:
Till, here meeting at night,
We shall have the delight
To discourse of our fortunes at leisure.

The day 's not too bright,
And the wind hits us right,
And all Nature does seem to invite us:
We have all things at will
For to second our skill;
As they all did conspire to delight us.

Or stream now, or still,
A large panier will fill!
Trout and grayling, to rise are so willing.
I dare venture to say,
''Twill be a bloody day;
And we all shall be weary of killing!'

Away, then, away!
We lose sport by delay!
But, first, leave our sorrows behind us!
If Misfortune do come;
We are all gone from home!
And a fishing she never can find us!

The Angler is free
From the cares that Degree
Finds itself with so often tormented:
And although we should slay
Each a hundred to-day;
'Tis a slaughter needs ne'er be repented!

And though we display
All our arts to betray
What were made for Man's pleasure and diet:
Yet both Princes and States
May, for all our quaint baits,
Rule themselves and their people in quiet.

We scratch not our pates;
Nor repine at the Rates
Our superiors impose on our living:
But do frankly submit;
Knowing they have more wit
In demanding, than we have in giving.

Whilst quiet we sit,
We conclude all things fit;
Acquiescing with hearty submission.
For, though simple, we know
That soft murmurs will grow,
At the last, unto downright sedition!

We care not, who says,
And intends it dispraise,
'That an Angler t' a Fool is next neighbour!'
Let him prate! What care we!
We're as honest as he!
And so, let him take that for his labour!

We covet no wealth
But the blessing of health;
And that greater, good conscience within!
Such devotion we bring
To our God and our King,
That from either no offers can win!

Whilst we sit and fish,
We do pray, as we wish,
For long life to our King JAMES the Second!
Honest Anglers then may,
Or they've very foul play,
With the best of good subjects be reckoned.

PRITHEE, little Boy! refrain!
      'Tis, in vain,
That thou at my heart dost aim!
For kind BACCHUS does so charm it;
      Nought but wine,
Nought but wine can ever warm it!

Tell me not of Ladies' eyes!
      I despise
All flames which from thence arise!
The highest Loves e'er yet created
      Are by wine,
Are by wine quenched, or abated!

I should Women tyrants find;
      If I whined,
When to me they proved unkind!
The first coldness I discover;
      I cure one,
I cure one heat by another!

After I my flame relate,
      If She hate,
I use her too, at that rate!
    For 'tis always my desire
      To do like,
To do like her I admire!

Therefore, though you were more fair
　　Than you are,
If unkind, I would not care!
Nothing more, or less, can move me
　　To love you,
To love you, than you to love me!

ADVICE FOR A LOVER.

CLEON! if you would have me true;
　Be you my precedent and guide!
Example sooner we pursue,
　Than the dull dictates of our pride.
　　Precepts of Virtue are too weak an aim;
　　'Tis Demonstration, that can best reclaim!

Shew me the path you'd have me go!
　With such a guide, I cannot stray!
What you approve, whate'er you do;
　It is but just I bend that way!
　　If true, my honour favours your design;
　　If false, revenge is the result of mine!

A Lover true, a Maid sincere,
　Are to be prized as things divine.
'Tis justice makes the blessing dear;
　Justice of love, without design.
　　And she that reigns not in a heart alone
　　Is never safe, nor easy, on her throne!

ONE night, when all the village slept,
    MYRTILLO's sad despair,
The wand'ring Shepherd waking kept,
    To tell the woods his care.

'Be gone!' said he, 'fond thought, be gone!
    Eyes, give your sorrows o'er!
Why should you waste your tears for one
    That thinks on you no more?

'Yet all the birds, the flocks, and Powers,
    That dwell within this grove,
Can tell how many tender hours
    We here have passed in love!

'Yon stars above (my cruel foes!)
    Have heard how She has sworn,
A thousand times, that, like to those
    Her flame should ever burn!

'But since She 's lost; O, let me have
    My wish, and quickly die!
In this cold bank, I'll make a grave;
    And there, for ever lie!

'Sad nightingales, the watch shall keep;
   And kindly here complain!'
Then down the Shepherd lay to sleep;
   But never waked again!

———

As AMORET with PHILLIS sat,
   One evening on the plain,
And saw the charming STREPHON wait
   To tell the Nymph his pain;

The threat'ning danger to remove,
   She whispered in her ear,
'Ah! PHILLIS! if you would not love;
   This Shepherd do not hear!

'None ever had so strange an art
   His Passion to convey.
Into a list'ning Virgin's heart;
   And steal her soul away!

'Fly! fly betimes! for fear you give
   Occasion for your fate!'
'In vain!' said she, 'in vain, I strive!
   Alas! 'tis now too late!'

I GRANT, a thousand oaths I swore,
　　I none would love but you!
But not to change would wrong me more,
　　Than breaking them would do!
Yet you thereby a truth will learn,
　　Of much more worth than I!
Which is, That Lovers which do swear,
　　Do also use to lie!

CLORIS does now possess that heart,
　　Which to you did belong;
But, though thereof she brags a while,
　　She shall not do so long!
She thinks, by being fair and kind,
　　To hinder my remove;
And ne'er so much as dreams, that Change
　　Above both those I love!

Then grieve not any more; nor think
　　My change is a disgrace!
For though it robs you of one slave;
　　It leaves another's place!
Which your bright eyes will soon subdue
　　With him does them first see:
For if they could not conquer more;
　　They ne'er had conquered me!

AN EPITAPH UPON

THOMAS, LORD FAIRFAX,

COMMANDER IN CHIEF FOR THE PARLIAMENT.

UNDER this stone does lie
One born for Victory.

FAIRFAX, the Valiant! and the only He
Who e'er, for that alone, a conqueror would be.
  Both sexes' virtues were in him combined:
  He had the fierceness of the manliest mind,
  And all the meekness too of womankind.
He never knew what Envy was, or Hate.
  His soul was filled with Worth and Honesty;
And with another thing, quite out of date,
      Called Modesty.

He ne'er seemed impudent, but in the Field: a place
Where Impudence itself dares seldom show its face.
  Had any stranger spied him in a room
  With some of those he had overcome,

And had not heard their talk; but only seen
      Their gestures and their mien:
They would have sworn he had, the vanquished been!
For, as they bragged, and dreadful would appear,
While they, their own ill luck in war repeated;
His modesty still made him blush to hear
      How often he had them defeated.

Through his whole life, the Part he bore
      Was wonderful and great:
And yet it so appeared in nothing more
      Than in his private last retreat.
      For 'tis a stranger thing to find
      One man of such a glorious mind,
      As can despise the Power he's got:
      Than millions of the Sots and Braves
      (Those despicable fools and knaves!);
            Who such a pudder make,
            Through dullness and mistake,
      In seeking after Power—and get it not!

When all the Nation he had won,
And with expense of blood had bought
      Store great enough, he thought,
      Of fame and of renown:
      He then, his Arms laid down

With full as little pride,
As if he had been of his Enemy's side;
Or one of them could do, that were undone!
He neither wealth, nor Places, sought!
For others, not himself, he fought!
He was content to know
(For he had found it so!),
That, when he pleased, to conquer he was able;
And leave the spoil and plunder to the rabble.
He might have been a King!
But yet he understood
How much it is a meaner thing
To be unjustly Great; than honourably Good!

This, from the World did admiration draw;
And from his friends, both love and awe:
Remembering what he did in fight before.
His foes lovèd him too!
As they were bound to do;
Because he was resolved to fight no more.
So, blessed of all, he died!
But far more blessed were we,
If we were sure to live, till we could see
A Man as great in War; as just in Peace, as he!

———

TO HIS MISTRESS.

WHAT a dull fool was I,
To think so gross a lie,
As that I ever was in love before!
I have, perhaps, known one or two
With whom I was content to be
At that, which they call 'keeping company.'
But, after all that they could do,
I still could be with more!
Their absence never made me shed a tear!
And I can truly swear,
That, till my eyes first gazed on you,,
I ne'er beheld that thing I could adore!

A world of things must curiously be sought,
A world of things must be together brought,
To make up charms which have the power to move, ⎫
Through a discerning eye, true love. ⎬
That is a masterpiece, above ⎭
What only looks and shape can do!
There must be Wit, and Judgement too!
Greatness of Thought, and Worth, which draw
From the whole World, respect and awe!

She that would raise a noble love, must find
Ways to beget a Passion for her mind!
She must be that, which She, to be would seem;
For all true love is grounded on esteem.

Plainness and Truth gain more a generous heart
Than all the crookèd subtleties of art.
   She must be (what said I ?), She must be *you*!
None but yourself that miracle can do!
At least, I'm sure, thus much I plainly see,
None but yourself e'er did it upon me!
'Tis you alone, that can my heart subdue!
To you alone, it always shall be true! . . .

———

   THOUGH, PHILLIS! your prevailing charms
Have forced me from my CELIA's arms,
That kind defence against all Powers
But those resistless eyes of yours;
Think not your conquest to maintain
By rigour and unjust disdain!
   In vain, fair Nymph! in vain you strive;
For Love does seldom Hope survive!
My heart may languish for a time;
Whilst all your glories, in their prime,
Can justify such cruelty,
By the same force that conquered me.
   When age shall come (at whose command
Those troops of beauties must disband!),
A tyrant's strength once took away;
What slave, so dull as to obey!
   [But if you will learn a nobler way
To keep this empire from decay,
And there, for ever, fix your throne;
Be kind, but kind to me alone!]

———

I.

SINCE CŒLIA 's my foe;
To a desert I'll go!
  Where some river
    For ever
Shall echo my woe!

The trees shall appear
More relenting than her;
  In the morning,
    Adorning
Each leaf with a tear.

When I make my sad moan
To the rocks, all alone;
  From each hollow
    Will follow
Some pitiful groan.

But, with silent disdain,
She requites all my pain!
  To my mourning,
    Returning
No answer again.

II.

Ah! CŒLIA! adieu!
When I cease to pursue,
　You'll discover
　　No Lover
Was ever so true!

Your sad Shepherd flies
From those dear cruel eyes!
　Which not seeing,
　　His being
Decays, and he dies!

Yet 'tis better to run
To the fate we can't shun;
　Than for ever
　　To strive for
What cannot be won!

What, ye Gods! have I done?
That AMYNTOR alone
　Is so treated
　　And hated
For loving but one!

THE NYMPH COMPLAINING

FOR THE DEATH OF HER FAWN.

THE wanton troopers, riding by,
Have shot my Fawn; and it will die!
Ungentle men! They cannot thrive,
To kill thee! Thou ne'er didst, alive,
Them any harm; alas, nor could
Thy death yet do them any good!
I'm sure I never wished them ill;
Nor do I, for all this! nor will!
But if my simple prayers may yet
Prevail with Heaven to forget
Thy murder; I will join my tears,
Rather than fail! But, O, my fears!
It cannot die so! Heaven's King
Keeps register of every thing;
And nothing may we use in vain!
E'en beasts must be with justice slain;
Else men are made their deodands!
Though they should wash their guilty hands
In this warm life-blood, which doth part
From thine, and wound me to the heart;
Yet could they not be clean! their stain
Is dyed in such a purple grain.
There is not such another in
The world, to offer for their sin!

Unconstant SYLVIO, when yet
I had not found him counterfeit,
One morning, (I remember well!)
Tied in this silver chain and bell,
Gave it to me. Nay! and I know
What he said then! I'm sure I do!
Said he, 'Look, how your huntsman here
Hath taught a Fawn to hunt his Dear!'
But SYLVIO soon had me beguiled.
This waxèd tame; while he grew wild!
And, quite regardless of my smart,
Left me his Fawn; but took his heart!

Thenceforth, I set myself to play
My solitary time away
With this; and very well content,
Could so mine idle life have spent!
For it was full of sport, and light
Of foot and heart; and did invite
Me to its game. It seemed to bless
Itself in me. How could I less
Than love it! O, I cannot be
Unkind t' a beast that loveth me!

Had it lived long, I do not know
Whether it too, might have done so
As SYLVIO did. His gifts might be,
Perhaps, as false, or more, than he!
But I am sure, for aught that I
Could, in so short a time, espy,

Thy love was far more better than
The love of false and cruel men!

   With sweetest milk and sugar, first
I, it at mine own fingers nurst;
And, as it grew, so every day
It waxed more white and sweet than they!
It had so sweet a breath! And oft
I blushed, to see its foot more soft
And white, shall I say, than my hand?
Nay! any Lady's in the land!

   It is a wondrous thing, how fleet .
'Twas on those little silver feet!
With what a pretty skipping grace,
It oft would challenge me the race!
And when 't had left me far away,
'Twould stay; and run again, and stay.
For it was nimbler much than hinds;
And trod as on the four winds!

   I have a garden of my own;
But so with roses overgrown,
And lilies, that you would it guess
To be a little wilderness:
And, all the Spring time of the year,
It only lovèd to be there.
   Among the beds of lilies, I
Have sought it oft, where it should lie:
Yet could not, till itself would rise,

Find it; although before mine eyes.
For, in the flaxen lilies' shade,
It like a bank of lilies laid.
 Upon the roses, it would feed
Until its lips e'en seemed to bleed!
And then to me 'twould boldly trip,
And print those roses on my lip.
 But all its chief delight was, still
On roses thus itself to fill;
And its pure virgin limbs to fold
In whitest sheets of lilies cold.
Had it lived long, it would have been
Lilies without! roses within!

 O, help! O, help! I see it faint!
And die as calmly as a Saint!
 See, how it weeps! The tears do come
Sad, slowly dropping like a gum.
So weeps the wounded balsam! So
The holy frankincense doth flow!
The brotherless HELIADES
Melt in such amber tears as these!
 I, in a golden vial, will
Keep these two crystal tears! and fill
It, till it do o'erflow, with mine!
Then place it in DIANA's shrine.

 Now my sweet Fawn is vanished to
Whither the swans and turtles go!
In fair Elysium to endure,

With milk-white lambs and ermines pure.
　O, do not run too fast! For I
Will but bespeak thy grave; and die!

　First, my unhappy statue shall
Be cut in marble; and withal,
Let it be weeping too! But there
Th' engraver, sure, his art may spare!
For I so truly thee bemoan,
That I shall weep, though I be stone,
Until my tears, still dropping, wear
My breast; themselves engraving there.
　There, at my feet, shalt thou be laid!
Of purest alabaster made;
For I would have thine image be
White, as I can; though not as thee!

AMETAS AND THESTYLIS

MAKING HAY-ROPES.

AMETAS. THINK'ST thou, that this love can stand,
　　　　Whilst thou still dost say me 'Nay!'?
　　Love unpaid does soon disband!
　　　　Love binds love, as hay binds hay!

THESTYLIS. Think'st thou, that this rope would twine,
　　　　If we both should turn one way?
　　Where both parties so combine,
　　　　Neither love will twist; nor hay!

AMETAS.	Thus, you vain excuses find;
	Which yourself and us delay!
	And Love ties a woman's mind
	Looser than with ropes of hay!

| THESTYLIS. | *What you cannot constant hope,* |
| | *Must be taken as you may!* |

| AMETAS. | Then let 's both lay by our rope; |
| | And go kiss within the hay! |

BERMUDAS.

WHERE the remote Bermudas ride
In th' Ocean's bosom unespied,
From a small boat, that rowed along,
The list'ning winds received this Song.

What should we do, but sing His praise
That led us, through the wat'ry maze,
Unto an isle so long unknown;
And yet far kinder than our own!
Where He, the huge sea-monsters wracks,
That lift the deep upon their backs;
He lands us on a grassy stage,
Safe from the storms' and Prelates' rage.

He gave us this eternal Spring,
Which here enamels every thing;

And sends the fowls to us, in care,
On daily visits through the air.
He hangs in shades, the orange bright,
Like golden lamps in a green night;
And does, in the pomegranates, 'close
Jewels more rich than Ormus shows!
　He makes the figs, our mouths to meet;
And throws the melons at our feet.
But apples, plants of such a price,
No tree could ever bear them twice!
With cedars, chosen by His hand
From Lebanon, He stores the land;
And makes the hollow seas, that roar,
Proclaim the ambergris on shore.

　He cast (of which we rather boast!)
The Gospel's Pearl upon our coast:
And, in these rocks, for us did frame
A Temple, where to sound His name.
　O, let our voice His praise exalt
'Till it arrive at heaven's vault!
Which, then, perhaps, rebounding, may
Echo beyond the Mexique Bay!

Thus sung they, in the English boat,
A holy and a cheerful note;
And, all the way, to guide their chime,
With falling oars, they kept the time.

———

THE GARDEN.

How vainly men themselves amaze
To win the palm! the oak! or bays!
And, their uncessant labours see
Crowned from some single herb, or tree;
Whose short, and narrow-vergèd, shade
Does prudently their toils upbraid!
While all flowers and all trees do close
To weave the garlands of repose.

Fair Quiet! have I found thee here;
And Innocence, thy sister dear?
Mistaken long, I sought you then
In busy companies of men.
Your sacred plants, if here below,
Only among the plants will grow!
Society is all but rude
To this delicious Solitude!

No white, nor red, was ever seen
So am'rous as this lovely green!
Fond Lovers, cruel as their flame,
Cut in these trees, their Mistress' name.
Little, alas, they know, or heed,
How far these beauties, hers exceed!
Fair trees! wheres'e'er your barks I wound;
No name shall but your own be found!

When we have run our Passion's heat;
Love hither makes his best retreat!
The Gods, that mortal beauty chase,
Still in a tree did end their race!
APOLLO hunted DAPHNE so,
Only that she might laurel grow!
And PAN did after SYRINX speed,
Not as a Nymph, but for a reed!

What wondrous life is this I lead!
Ripe apples drop about my head!
The luscious clusters of the vine,
Unto my mouth do crush their wine!
The nectarine, and curious peach,
Into my hands themselves do reach!
Stumbling on melons, as I pass;
Insnared with flowers, I fall on grass!

Meanwhile, the mind, from pleasure less,
Withdraws into its happiness!
The mind, that Ocean where each kind
Does straight its own resemblance find:
Yet it creates, transcending these,
Far other worlds, and other seas;
Annihilating all that 's made
To a green thought in a green shade!

Here, at the fountain's sliding foot,
Or at some fruit-tree's mossy root;
Casting the body's vest aside,
My soul, into the boughs does glide!
There, like a bird, it sits and sings!
Then whets, and combs, its silver wings!
And (till prepared for longer flight)
Waves in its plumes, the various light!

Such was that happy garden state,
While Man there walked without a mate.
After a place so pure and sweet,
What other Help could yet be meet!
But 'twas beyond a mortal's share
To wander solitary there!
Two Paradises 'twere in one,
To live in Paradise alone!

How well the skilful Gard'ner drew,
Of flowers and herbs, this Dial new!
Where, from above, the milder sun
Does through a fragrant Zodiac run:
And, as it works, th' industrious bee
Computes its time, as well as we.
How could such sweet and wholesome hours
Be reckoned, but with herbs and flowers!

A DIALOGUE

BETWEEN THE RESOLVED SOUL AND CREATED PLEASURE.

COURAGE, my Soul! now learn to wield
The weight of thine immortal shield!
Close on thy head, thy helmet bright!
Balance thy sword against the fight!
    See, where an army, strong as fair,
With silken banners spreads the air!
Now, if thou be'st that thing divine,
In this day's combat, let it shine!
And show, that Nature wants an art
To conquer one resolvèd heart.

PLEASURE. Welcome, the Creation's guest!
        Lord of Earth, and Heaven's heir!
    Lay aside that warlike crest;
        And of Nature's banquet share!
            Where the souls of fruits and flowers
            Stand prepared to heighten yours.

SOUL.   I sup above; and cannot stay
        To bait so long upon the way.

PLEASURE. On these downy pillows lie!
        Whose soft plumes will thither fly.

On these roses! strowed so plain,
Lest one leaf thy side should strain!

SOUL.

My gentler rest is on a thought;
Conscious of doing what I ought.

PLEASURE.

If thou be'st with perfumes pleased,
Such as oft the Gods appeased;
Thou, in fragrant clouds, shalt show
Like another God below!

SOUL.

A Soul that knows not to presume,
Is Heaven's, and its own, perfume!

PLEASURE.

Every thing does seem to vie
Which should first attract thine eye;
But since none deserves that grace,
In this crystal, view thy face!

SOUL.

When the Creator's skill is prized;
The rest is all but earth disguised!

PLEASURE.

Hark! how Music then prepares
For thy stay, these charming Airs!
Which, the posting winds recall;
And suspend the river's fall.

SOUL.

Had I but any time to lose;
On this, I would it all dispose!
Cease, Tempter! None can chain a mind;
Whom this sweet chordage cannot bind!

87

CHORUS.
Earth cannot shew so brave a sight,
As when a single Soul does fence
The batt'ry of alluring Sense!
And Heaven views it with delight!
Then, persevere! For still new Charges
sound;
And if thou overcom'st, thou shalt be
crowned!

PLEASURE. All this fair, and cost, and sweet,
Which scatt'ringly doth shine,
Shall within one Beauty meet;
And She be only thine!

SOUL. If things of sight such heavens be;
What Heavens are those, we cannot see!

PLEASURE. Wheresoe'er thy foot shall go,
The minted gold shall lie;
Till thou purchase all below,
And want new worlds to buy!

SOUL. Were 't not a price; who'ld value gold!
And that 's worth nought, that can be sold!

PLEASURE. Wilt thou all the Glory have
That War, or Peace, commend?
Half the World shall be thy slave;
The other half, thy friend!

SOUL.   What friends! if to myself untrue!
        What slaves! unless I captive you!

PLEASURE. Thou shalt know each Hidden Cause,
        And see the Future Time!
        Try what depth the Centre draws;
        And then to Heaven climb!

SOUL.   None thither mounts by the degree
        Of Knowledge; but, Humility!

CHORUS. *Triumph! triumph! victorious Soul!*
        The World has not one Pleasure more!
        The rest does lie beyond the Pole;
        And is thine everlasting store!

THE PICTURE OF LITTLE T. C.

IN A PROSPECT OF FLOWERS.

SEE, with what simplicity
This Nymph begins her golden days!
  In the green grass She loves to lie!
    And there, with her fair aspect, tames
    The wilder flowers, and gives them names:
But only with the roses plays;
    And them does tell,
What colour best becomes them, and what smell!

89

Who can foretell, for what high cause
This Darling of the Gods was born!
Yet this is She, whose chaster laws
    The wanton Love shall one day fear!
    And, under her command severe,
See his bow broke; and ensigns torn.
        Happy, who can
Appease this virtuous enemy of Man!

O, then, let me, in time, compound
And parley with those conquering eyes!
    Ere they have tried their force to wound!
    Ere, with their glancing wheels, they drive
    In triumph over hearts that strive;
And them that yield, but more despise!
        Let me be laid
Where I may see thy glories, from some shade!

Meantime, whilst every verdant thing
Itself does at thy beauty charm,
    Reform the errors of the Spring!
    Make that the Tulips may have share
    Of sweetness; seeing they are fair!
And Roses, of their thorns disarm!
        But most procure
That Violets may, a longer age endure!

90

But, O, young Beauty of the Woods!
Whom Nature courts with fruits and flowers,
Gather the flowers; but spare the buds!
Lest FLORA, angry at thy crime,
To kill her infants in their prime,
Do quickly make th' example yours!
And, ere we see,
Nip in the blossom all our hopes, and thee!

THE FAIR SINGER.

To make a final conquest of all me,
LOVE did compose so sweet an Enemy,
In whom both beauties, to my death agree;
Joining themselves in fatal harmony.
That, while She, with her Eyes my heart does bind;
She, with her Voice, might captivate my mind!

I could have fled from One but singly fair!
My disentangled Soul itself might save,
Breaking the curlèd trammels of her hair!
But how should I avoid to be her slave;
Whose subtle art invisibly can wreathe
My fetters, of the very air I breathe!

It had been easy fighting, in some plain,
Where victory might hang in equal choice:
But all resistance against her is vain;
Who has th' advantage both of Eyes and Voice!
And all my forces needs must be undone;
She having gainèd both the Wind and Sun!

TO HIS COY MISTRESS.

HAD we but world enough, and time,
This coyness, Lady! were no crime!
We would sit down, and think which way
To walk, and pass our long Love's Day.
Thou, by the Indian Ganges' side,
Shouldst rubies find; I, by the tide
Of Humber would complain! I would
Love you ten years before the Flood;
And you should, if you please, refuse
Till the Conversion of the Jews!
My vegetable love should grow
Vaster than Empires, and more slow!
A hundred years should go to praise
Thine eyes, and on thy forehead gaze!
Two hundred, to adore each breast:
But thirty thousand to the rest!
An Age, at least, to every part;
And the Last Age should show your heart!
For, Lady! you deserve this State;
Nor would I love at lower rate!

But, at my back, I always hear
Time's wingèd Chariot hurrying near!

And yonder, all before us, lie
Deserts of vast Eternity!
Thy beauty shall no more be found;
Nor, in thy marble vault, shall sound
My echoing Song! Then, worms shall try
That long preserved virginity!
And your quaint honour turn to dust;
And into ashes, all my lust!
The grave 's a fine and private place!
But none, I think, do there embrace.

Now, therefore, while the youthful hue
Sits on thy skin like morning dew,
And while thy willing soul transpires
At every pore, with instant fires;
Now, let us sport us, while we may!
And now, like am'rous birds of prey,
Rather, at once, our time devour;
Than languish in his slow chapt power!
Let us roll all our strength, and all
Our sweetness, up into one ball;
And tear our pleasures, with rough strife,
Through the Iron Gates of Life!
Thus, though we cannot make our sun
Stand still; yet we will make him run!

———

EYES AND TEARS.

How wisely Nature did decree,
With the same Eyes, to weep and see!
That, having viewed the object vain,
They might be ready to complain!

And since the self-deluding Sight
In a false angle takes each height;
These Tears, which better measure all,
Like wat'ry lines and plummets fall!

Two Tears (which Sorrow long did weigh
Within the scales of either Eye;
And then paid out in equal poise)
Are the true price of all my joys!

What in the world most fair appears,
Yea, even Laughter, turns to Tears!
And all the jewels which we prize,
Melt in these Pendants of the Eyes!

I have through every garden been,
Amongst the red, the white, the green;
And yet, from all the flowers I saw,
No honey, but these Tears could draw!
94

So the all-seeing sun, each day,
Distils the world with chemic ray:
But finds the essence only showers;
Which, straight, in pity back he pours!

Yet happy they, whom Grief doth bless!
That weep the more, and see the less!
And, to preserve their sight more true,
Bathe still their Eyes in their own dew!

So MAGDALEN, in Tears more wise,
Dissolved those captivating Eyes;
Whose liquid chains could flowing meet
To fetter her Redeemer's feet. . . .

The sparkling glance that shoots desire,
Drenched in these waves, does lose its fire!
Yea, oft the Thund'rer pity takes,
And here the hissing lightning slakes!

The incense was to Heaven so dear;
Not as a perfume, but a Tear!
And stars show lovely in the night,
But as they seem the Tears of Light!

Ope then, mine Eyes! your double sluice;
And practise so your noblest use!
For others too can see, or sleep;
But only Human Eyes can weep!

Now, like two clouds, dissolving, drop;
And at each Tear, in distance stop!
Now, like two fountains, trickle down!
Now, like two floods, o'erturn and drown!

Thus, let your streams o'erflow your springs,
Till Eyes and Tears be the same things;
And each, the other's difference bears;
These weeping Eyes! those seeing Tears!

THE MOWER TO THE GLOWWORMS

Ye living Lamps! by whose dear light,
    The nightingale does sit so late;
And, studying all the summer night,
    Her matchless Songs does meditate.

Ye country Comets! that portend
    No war, nor Prince's funeral!
Shining unto no higher end
    Than to presage the grasses' fall!

Ye Glowworms! whose officious flame
    To wand'ring Mowers shows the way;
That, in the night, have lost their aim,
    And after foolish fires do stray.

Your courteous lights, in vain you waste;
    Since JULIANA here is come!
For she, my mind hath so displaced,
    That I shall never find my home!

THE GALLERY.

CLORA, come view my Soul! and tell,
Whether I have contrived it well?
Now all its several lodgings be
Composed into one Gallery:
And the great arras-hangings, made
Of various faces, by are laid;
That, for all furniture, you'll find
Only your picture in my mind!

Here, thou art painted in the dress
Of an inhuman Murderess!
Examining upon our hearts,
Thy fertile shop of cruel arts,
Engines more keen than ever yet
Adornèd tyrant's cabinet!
Of which, the most tormenting are
Black Eyes, red Lips, and curlèd Hair.

But, on the other side, th' art drawn
Like to AURORA in the dawn!
When, in the East, she slumb'ring lies,
And stretches out her milky thighs:
While all the morning Quire does sing,
And manna falls, and roses spring;
And, at thy feet, the wooing doves
Sit perfecting their harmless loves.

Like an Enchantress here thou show'st!
Vexing thy restless Lover's ghost;
And, by a light obscure, dost rave
Over his entrails, in the cave:
Divining thence, with horrid care,
How long thou shalt continue fair!
And, when informed, them throw'st away!
To be the greedy vulture's prey.

But, against that, thou sitt'st afloat,
Like VENUS in her pearly boat!
The halcyons, calming all that 's nigh,
Betwixt the air and water fly:
Or if some rolling wave appears,
A mass of ambergris it bears;
Nor blows more wind than what may well
Convoy the perfume to the smell.

These pictures, and a thousand more,
Of thee, my Gallery does store;
In all the forms thou canst invent,
Either to please me, or torment!
For thou alone, to people me,
Art grown a num'rous colony!
And a Collection choicer far
Than, or Whitehall's, or Mantua's, were.

But of these pictures, and the rest;
That at the Entrance likes me best!
Where the same posture, and the look,
Remain with which I first was took.
A tender Shepherdess, whose hair
Hangs loosely playing in the air,
Transplanting flowers from the green hill,
To crown her head, and bosom fill.

— —

ON A DROP OF DEW.

SEE, how the orient dew,
Shed from the bosom of the Morn
    Into the blowing roses,
Yet careless of its mansion new,
For the clear region, where 'twas born,
    Round in itself incloses!
And, in its little globe's extent,
Frames, as it can, its native element.

How it, the purple flower does slight!
    Scarce touching where it lies;
But gazing back upon the skies,
    Shines with a mournful light,
        Like its own tear;
Because so long divided from the Sphere.

Restless it rolls, and unsecure,
　　Trembling lest it grow impure,
　　Till the warm sun pities its pain;
And to the skies exhales it back again.

　　So the Soul, that drop, that ray
Of the clear Fountain of Eternal Day,
Could it, within the human flower be seen,
　　Rememb'ring still its former height,
　　Shuns the sweet leaves and blossoms green;
　　And, re-collecting its own light,
Does, in its pure and circling thoughts, express
The greater Heaven in a heaven less.

　　In how coy a figure wound,
　　　Every way it turns away!
　　So the World excluding round;
　　　Yet receiving in the Day!
　　Dark beneath; but bright above!
　　Here disdaining; there in love!
How loose, and easy hence to go!
　　How girt, and ready to ascend!
Moving but on a point below,
　　It all about does upward bend!

Such did the Manna's sacred dew distil
White and entire, though congealed and chill:
Congealed on earth; but does, dissolving, run
Into the glories of th' Almighty Sun!

WITH amorous wiles and perjured eyes,
    False DAMON did me move!
Like charming winds, his kindling sighs
    First fanned me into love!
My thriving Passion he did feed,
    Whilst it was young and slight;
But, ah! when there was greatest need,
    Alas! he starves it quite!

Was ever more injustice known?
    O, DAMON! prithee, say!
To fit my heart for thee alone;
    And cast it now away!
Henceforth, my Passion I shall hate!
    'Cause it gained none for me:
Yet love it too, (such is my fate!)
    Because it was for thee!

Thy heart, I never will upbraid!
    Although it mine did kill.
Ah! think upon an injured Maid,
    That's forced to love thee still!
But Justice may the tables turn,
    In vindicating me;
And thou, with equal torments burn
    For one who loves not thee!

A PASTORAL SONG.

As I was sitting on the grass
  Within a silent shady grove,
I overheard a Country Lass,
  Was there bewailing of her Love.
      ' My Love,' says she,
      ' Is ta'en from me ;
And to the Wars is prest and gone.
      He 's marched away,
      And gone to sea ;
Alack ! alack ! and a welladay !
  And left me here alone.

' My Love, he was the kindest man ;
  There 's none that 's like him, in the town !
He gently takes me by the hand ;
  And gave me many a green gown !
      With kisses sweet
      He would me treat,
And often sing a Roundelay ;
      And sometimes smile,
      Then chat a while ;
That so we might the time beguile
  A life-long summer's day !

' My Love, on May Day, still would be
  The earliest up of all the rest !
With scarves and ribbons, then would he,
  Of all the crew, be finest drest !

With Morris bells
And fine things else :
But when the pipe began to play,
He danced so well,
I heard them tell,
That he did, all the rest excel ;
And bore the bell away.

'The man, that took my Love away,
Was too too harsh, and too severe.
I gently, on my knees, did pray
That he, my Love would then forbear !
I offered too
A breeding ewe
And chilver-lamb, that were my own.
Do what I could,
It did no good !
He left me, in this pensive mood,
To sigh, and make my moan !'

———

[The following Poem is imitated from the one by HERRICK at VI. 129.]

Go, perjured Maid ! to all extremes inclined !
First, so endearing ; after, so unkind !
As cruel, as inconstant, is thy mind !
Go to my Rival ! Leave me to complain !
Tell him from me, He has not long to reign !
I know your heart. You'll quickly change again !

———

TRUE LOVE REQUITED,

OR

THE BAILIFF'S DAUGHTER OF ISLINGTON.

THE YOUNG MAN'S FRIENDS, THE MAID DID SCORN;
'CAUSE SHE WAS POOR, AND LEFT FORLORN.
THEY SENT THE ESQUIRE TO LONDON FAIR,
TO BE AN APPRENTICE SEVEN YEAR.
AND WHEN HE OUT ON 'S TIME WAS COME,
HE MET HIS LOVE A GOING HOME:
AND THEN, TO END ALL FURTHER STRIFE,
HE TOOK THE MAID TO BE HIS WIFE.

There was a Youth, and a well-belovèd Youth,
   And he was a 'Squire's son.
He loved the Bailiff's daughter dear,
   That lived in Islington.

But she was coy; and she would not believe
   That he did love her so;
No! nor at any time she would
   Any countenance to him show.

But when his friends did understand
   His fond and foolish mind;
They sent him up to fair London,
   An Apprentice for to bind.

And when he had been seven long years,
  And his Love he did not see,
'Many a tear have I shed, for her sake;
  When she little thought of me!'

All the Maids of Islington
  Went forth to sport and play;
All but the Bailiff's daughter dear.
  She secretly stole away.

Then she put off her gown of gray;
  And put on her puggish attire.
She is up to fair London gone,
  Her True Love to require.

Now as she went along the road,
  The weather being hot and dry,
There, was she aware of her True Love,
  At length, came riding by.

She stepped to him, as red as any rose,
  And took him by the bridle-ring,
'I pray you, kind Sir, give me one penny
  To ease my weary limb!'

'I prithee, sweet Heart! canst thou tell me,
  Where that thou wast born?'
'At Islington, kind Sir!' said she,
  'Where I have had many a scorn!'

'I prithee, sweet Heart! canst thou tell me,
  Whether thou dost know
The Bailiff's daughter of Islington?'
  'She is dead, Sir! long ago.'

'Then will I sell my goodly steed,
  My saddle, and my bow;
I will into some far country,
  Where no man doth me know.'

'O, stay! O, stay! thou goodly Youth!
  Here she standeth by thy side!
She is alive! she is not dead;
  And is ready to be thy Bride!

'O, farewell, Grief! and welcome, Joy
  Ten thousand times and more!
For now I have seen my own True Love,
  That I thought I should have seen no more!'

John Oldham.

THE CUP.

A PARAPHRASE OF ANACREON.

[See page 113.]

MAKE me a bowl, a mighty bowl,
Large as my capacious soul!
Vast as my thirst is! Let it have
Depth enough to be my grave!
I mean, the grave of all my care;
For I intend to bury 't there.
Let it of silver fashioned be,
Worthy of wine! worthy of me!
Worthy to adorn the Spheres,
As that bright Cup among the stars! . . .

Yet draw no shapes of armour there,
No casque, nor shield, nor sword, nor spear,
Nor Wars of Thebes, nor Wars of Troy,
Nor any other martial toy.
For what do I, vain armour prize!
Who mind not such rough exercise;
But gentler sieges, softer wars,
Fights that cause no wounds, or scars.
I'll have no battles on my plate;
Lest sight of them should brawls create!
Lest that provoke to quarrels too;
Which wine itself enough can do. . . .

THE COMMONS' PETITION
TO KING CHARLES II.

In all humanity, we crave
Our Sovereign may be our slave!
And humbly beg, that he may be
Betrayed by us most loyally!
And if he please once to lay down
His sceptre, dignity, and crown;
We'll make him, for the time to come,
The greatest Prince in Christendom!

THE KING'S ANSWER.

Charles, at this time, having no need,
Thanks you as much as if he did.

———

THE KING'S EPITAPH.

Here lies a great and mighty King;
  Whose promise none relied on!
He never said a foolish thing;
  Nor ever did a wise one!

———

PLAIN DEALING'S DOWNFALL.

LONG time, PLAIN DEALING, in the haughty Town,
Wand'ring about, though in thread-bare gown,
At last, unanimously, was cried down.

When, almost starved, she to the Country fled;
In hopes, though meanly, she should there be fed,
And tumble nightly on a pea-straw bed.

But KNAV'RY, knowing her intent, took post,
And rumoured her approach through every coast;
Vowing his ruin, that should be her host.

Frighted at this, each rustic shut his door,
Bid her, Be gone, and trouble him no more!
For he that entertained her, must be poor!

At this, grief seized her! grief too great to tell!
When, weeping, sighing, fainting, down she fell!
Whilst KNAV'RY, laughing, rung her Passing Bell.

A PASTORAL DIALOGUE
BETWEEN ALEXIS AND STREPHON.

ALEXIS.　THERE sighs not on the plain
　　　　So lost a Swain as I!
　　　Scorched up with love, frozen with disdain;
　　　Of killing sweetness I complain!

STREPHON.　If 'tis CORINNA, die!
　　　Since first my dazzled eyes were thrown
　　　　On that bewitching face,
　　　Like ruined birds robbed of their young,
　　　Lamenting, frighted, and undone,
　　　　I fly from place to place!
　　　Framed by some cruel Powers above,
　　　　So nice she is, and fair;
　　　None from undoing can remove!
　　　Since all who are not blind, must love!
　　　Who are not vain, despair!

ALEXIS.　The Gods no sooner give a grace;
　　　　But, fond of their own art,
　　　Severely jealous, ever place,
　　　To guard the glories of a face,
　　　　A dragon in the heart.

Proud and ill-natured Powers they are!
  Who, peevish to mankind,
For their own honour's sake, with care
Make a sweet Form divinely fair;
  Then add a cruel mind!

STREPHON. Since she's insensible of Love,
  By Honour taught to hate:
If we, forced by decrees above,
Must sensible to Beauty prove;
  How tyrannous is Fate!

ALEXIS. I to the Nymph have never named
  The cause of all my pain!

STREPHON. Such bashfulness may well be blamed!
For since to serve we're not ashamed;
  Why should she blush to reign?

ALEXIS. But if her haughty heart despise
  My humble proffered one;
The just compassion she denies,
I may obtain from others' eyes!
  Hers are not fair alone!
Devouring flames require new food;
  My heart's consumed almost!
New fires must kindle in her blood;
Or mine go out! and that's as good!

STREPHON.   Wouldst live, when Love is lost?
                    Be dead, before thy Passion dies!
                       For if thou shouldst survive,
                    What anguish would the heart surprise,
                    To see her flames begin to rise;
                       And thine no more alive!

ALEXIS.   Rather, what pleasure should I meet
                       In my triumphant scorn!
                    To see my tyrant at my feet;
                    While, taught by her, unmoved I sit,
                       A tyrant in my turn!

STREPHON.   Ungentle Shepherd! cease, for shame!
                       Which way can you pretend
                    To merit so divine a flame;
                    Who to dull life make a mean claim,
                       When Love is at an end?
                    As trees are by their bark embraced;
                       Love to my soul doth cling!
                    When, torn by the herd's greedy taste,
                    The injured plants feel they're defaced,
                       They wither in the Spring!
                    My rifled love would soon retire,
                       Dissolving into air;
                    Should I, that Nymph cease to admire:
                    Blessed in whose arms, I will expire!
                       Or at her feet, despair!

———

UPON HIS DRINKING IN A BOWL.

[See page 107.]

VULCAN! contrive me such a Cup
    As NESTOR used of old!
Shew all thy skill to trim it up!
    Damask it round with gold!

Make it so large that, filled with Sack
    Up to the swelling brim,
Vast Toasts, on the delicious Lake,
    Like ships at sea, may swim!

Engrave not battle on his cheek;
    With war, I've nought to do!
I'm none of those that took Maastricht;
    Nor Yarmouth leaguer knew!

Let it no name of Planets tell,
    Fixed Stars, or Constellations:
For I am no Sir SIDROPHEL,
    Nor none of his relations!

But carve thereon a spreading vine;
    Then add two lovely boys!
Their limbs in amorous folds intwine,
    The type of future joys!

CUPID and BACCHUS, my Saints are!
  May Drink and Love still reign!
With Wine I wash away my cares;
  And then to Love again!

———

WHILE in divine PANTHEA's charming eyes,
I view the naked Boy, that basking lies,
I grow a God! so blest, so blest am I
With sacred rapture, and immortal joy!

But absent, if she shines no more,
And hides the suns that I adore;
Straight, like a wretch, despairing I
Sigh, languish in the shade, and die!

O, I were lost in endless night,
If her bright presence brought not light!
Then I revive, blest as before!
The Gods themselves can be no more!

———

  PITY, fair SAPPHO! one that dies
A victim to your beauteous eyes!
For while on them I dare to gaze,
Their dazzling glories so amaze,
My soul does melt with new desire!
I rave! I burn with secret fire!
And, blessing the dear Cause, expire!

To this moment, a rebel; I throw down my Arms,
Great LOVE! at first sight of OLINDA's bright charms!
Made proud, and secure, by such forces as these;
You may now play the tyrant as soon as you please!

When Innocence, Beauty, and Wit, do conspire
To betray, and engage, and inflame, my desire;
Why should I decline what I cannot avoid;
And let pleasing hope, by base fear be destroyed?

Her Innocence cannot contrive to undo me!
Her Beauty 's inclined, or why should it pursue me?
And Wit has, to Pleasure been ever a friend;   [end!
Then what room for Despair? since Delight is Love's

There can be no danger in Sweetness and Youth;
Where Love is secured by Good Nature and Truth!
On her Beauty I'll gaze, and of Pleasure complain;
While ev'ry kind look adds a link to my chain!

'Tis more to maintain, than it was to surprise!
But her Wit leads in triumph the slave of her Eyes!
I beheld, with the loss of my freedom before;
But, hearing, for ever must serve and adore!

Too bright is my Goddess! Her Temple too weak!
Retire, Divine Image! I feel my heart break!
Help, LOVE! I dissolve in a rapture of charms,
At the thought of those joys, I should meet in her arms!

John Wilmot, Earl of Rochester

My dear Mistress has a heart
  Soft as those kind looks She gave me,
When with Love's resistless Art
  And her eyes She did enslave me.
But her constancy's so weak,
  She's so wild, and apt to wander:
That my jealous heart would break,
  Should we live one day asunder!

Melting joys about her move,
  Killing pleasures, wounding blisses
She can dress her eyes in Love:
  And her lips can arm with kisses
Angels listen when She speaks:
  She's my delight, all mankind's wonder:
But my jealous heart would break,
  Should we live one day asunder

———

    I promised Sylvia to be true:
Nay swore it and I swore it too,
And that she might believe me more,
Gave her a writing what I swore
    Not valuing her smiles or Lovers kind,
So long as Heaven so long they're kind!
Twas on a leaf. The wind but blew:
Away went Leaf and Promise too.
        &c

———

WOMAN'S HONOUR.

Love bade me hope; and I obeyed.
  Phillis continued still unkind.
'Then you may e'en despair!' he said,
  'In vain, I strive to change her mind!

'Honour 's got in; and keeps her heart!
  Durst he but venture once abroad;
In my own right, I'd take your part,
  And shew myself the mightier God!

'This huffing Honour domineers
  In breast alone, where he has place;
But if true gen'rous Love appears,
  The Hector dares not show his face!'

Let me still languish, and complain,
  Be most inhumanly denied;
I have some pleasure, in my pain!
  She can have none, with all her pride!

I fall a sacrifice to Love!
  She lives a wretch, for Honour's sake!
Whose tyrant does most cruel prove?
  The difference is not hard to make!

Consider, real Honour then!
  You'll find hers cannot be the same!
'Tis noble confidence, in Men;
  In Woman, mean distrustful shame!

———

INSULTING Beauty! you misspend
  Those frowns upon your slave!
Your scorn, against such rebels bend,
Who dare with confidence pretend
That other eyes, their hearts defend
  From all the charms you have!

Your conquering eyes so partial are,
  Or mankind is so dull;
That, while I languish in despair,
Many proud senseless hearts declare,
They find you not so killing fair,
  To wish you merciful!

They, an inglorious freedom boast;
  I triumph in my chain!
Nor am I unrevenged, though lost,
Nor you unpunished, though unjust,
When I alone, who love you most,
  Am killed by your disdain!

———

CAIUS CORNELIUS GALLUS

IMITATED.

[See IV. 16.]

My Goddess, Lydia, heavenly fair!
As lilies sweet, as soft as air!
Let loose thy tresses, spread thy charms;
And to my Love give fresh alarms!

O, let me gaze on those bright eyes;
Though sacred lightning from them flies!
Shew me that soft, that modest, grace;
Which paints with charming red thy face!

Give me ambrosia in a kiss;
That I may rival Jove in bliss!
That I may mix my soul with thine;
And make the pleasure all divine!

O, hide thy bosom's killing white!
(The Milky Way is not so bright!)
Lest you my ravished soul oppress
With Beauty's pomp, and sweet excess!

Why draw'st thou from the purple flood
Of my kind heart, the vital blood?
Thou art, all over, endless charms!
O, take me, dying, to thy arms!

GIVE me leave to rail at you!
I ask nothing but my due!
To call you false! and then to say,
'You shall not keep my heart a day!'
But, alas! against my will,
I must be your captive still!
Ah! be kinder then; for I
Cannot change, and would not die!

Kindness has resistless charms!
    All besides but meekly move!
Fiercest Anger it disarms;
    And clips the wings of flying Love!
Beauty does the heart invade;
Kindness only can persuade!
It gilds the Lover's servile chain;
And makes the slave grow pleased again!

THE ANSWER.

NOTHING adds to your fond fire
    More than scorn and cold disdain!
I, to cherish your desire,
    Kindness used; but 'twas in vain!
You insulted on your slave!
    Humble love, you soon refused!
Hope not then a power to have,
    Which ingloriously you used!

Think not, THIRSIS! I will e'er,
 By my love, my empire lose!
You grow constant through despair!
 Love returned, you would abuse!
Though you still possess my heart;
 Scorn and rigour I must feign!
Ah! forgive that only art
 Love has left, your love to gain!

You, that could my heart subdue,
 To new conquests ne'er pretend!
Let your example make me true;
 And of a conquered foe, a friend!
Then, if e'er I should complain
Of your empire, or my chain,
Summon all your powerful charms,
And kill the rebel in your arms!

———

CONSTANCY.

I CANNOT change, as others do,
 Though you unjustly scorn;
Since that poor Swain, that sighs for you,
 For you alone was born!
No! PHILLIS! No! your heart to move,
 A surer way I'll try!
And to revenge my slighted love,
Will still love on! will still love on, and die!

When killed with grief AMYNTAS lies;
   And you to mind shall call
The sighs, that now unpitied rise;
   The tears, that vainly fall:
That welcome hour, that ends this smart,
   Will then begin your pain!
For such a faithful, tender heart
Can never break, can never break in vain!

———

PHILLIS! be gentler! I advise;
   Make up for time misspent!
When Beauty on its death-bed lies,
   'Tis high time to repent!

Such is the malice of your fate,
   That makes you old so soon;
Your pleasure ever comes too late,
   How early e'er begun!

Think, what a wretched thing is She;
   Whose stars contrive, in spite,
The morning of her Love should be
   Her fading Beauty's night! . . .

———

LOVE AND LIFE.

ALL my Past Life is mine no more!
   The flying hours are gone,
Like transitory dreams given o'er;
Whose images are kept in store
   By memory alone!

The time that is to come is not!
   How can it then be mine?
The Present Moment 's all my lot,
And that, as fast as it is got,
   PHILLIS! is only thine!

Then, talk not of inconstancy,
   False hearts, and broken vows!
If I, by miracle, can be
This life-long minute true to thee;
   'Tis all that Heaven allows!

A FAREWELL TO LOVE.

Once more, Love's mighty chains are broke!
  His strength and cunning I defy!
Once more, I have thrown off his yoke,
  And am a Man; and do despise the Boy:
Thanks to her pride, and her disdain,
  And all the follies of a scornful mind.
I had ne'er possessed my heart again,
  If fair Miranda had been kind!

Welcome, fond wanderer! as ease
  And plenty to a wretch in pain,
That (worn with want and a disease)
  Enjoys his health, and all his friends again!
Let others waste their time and youth,
  Watch and look pale, to gain a peevish Maid;
And learn, too late, this dear-bought truth,
  At length, they're sure to be betrayed!

I NEVER saw a face till now,
   That could my Passion move!
I liked, and ventured many a vow;
   But durst not think of Love!

Till Beauty, charming every sense,
   An easy conquest made;
And shewed the vainness of defence,
   Where PHYLLIS does invade.

But O, her colder heart denies
   The thoughts her looks inspire:
And while, in ice, that frozen lies;
   Her eyes dart only fire!

Betwixt extremes, I am undone!
   Like plants too northward set;
Burnt by too violent a sun,
   Or chilled for want of heat.

THE INQUEST.

WHERE 's absent CLELIA?
    Where are those eyes,
    That steal away
    My heart in play;
And over it so strangely tyrannize?

I thought I had been free:
    But looking round,
    Alas, for me!
    I nought could see;
Yet found myself in fetters closely bound!

I laid me down to rest:
    And yet my mind
    Was still opprest;
    And in my breast,
I did a hundred thousand torments find!

I walked the City round,
    In search of ease:
    But nothing found
    On which to ground
A hope of remedy for my disease.

Into the Country, straight
 I made repair,
 To mitigate
 My cruel fate;
But I found nothing there but sad Despair.

I viewed the archèd sky,
 And foaming sea.
 The first too high
 For me to fly;
And t' other deep, as is my misery!

I could not tell what course,
 Or way, to steer!
 Or by what force
 To gain remorse;
And ease my heart of this, my cruel fear!

At last, my CLELIA came!
 O, blest reprieve!
 And ceased to blame
 My ardent flame;
And, for her sake, commanded me to live!

What happiness was this,
 To one as lost!
 O, who could wish,
 So great a bliss!
Half starved at sea, to gain so blest a coast!

THE OLD MAN'S WISH.

THE OLD MAN, HE DOTH WISH FOR WEALTH IN VAIN;
  BUT HE DOTH NOT THE TREASURE GAIN:
FOR IF, WITH WISHES HE THE SAME COULD HAVE,
  HE WOULD NOT MIND, NOR THINK UPON, THE GRAVE!

IF I live to grow old (for I find I go down!),
Let this be my fate in a country town!
Let me have a warm house, with a stone at the gate;
And a cleanly young girl to rub my bald pate!
    May I govern my Passion with an absolute sway!
    And grow wiser and better, as my strength wears away,
    Without gout or stone, by a gentle decay!

In a country town, by a murmuring brook,
The ocean at distance, on which I may look;
With a spacious plain, without hedge or stile,
And an easy pad-nag to ride out a mile;
    May I govern my Passion with an absolute sway! &c.

With a pudding on Sunday, and stout humming liquor,
And remnants of Latin to puzzle the Vicar;
With a hidden reserve of Burgundy wine,
To drink the King's Health as oft as I dine;
    May I govern my Passion with an absolute sway! &c.

With PLUTARCH, and HORACE, and one or two more
Of the best Wits that lived in the Ages before;
With a dish of roast mutton, not venison, nor teal,
And clean, though coarse, linen at every meal;
    May I govern my Passion with an absolute sway! &c.

Walter Pope, M.D.

And if I should have guests, I must add to my Wish,
On Fridays, a mess of good buttered fish!
For, full well I do know, and the truth I reveal,
I had better do so, than come short of a meal!
    May I govern my Passion with an absolute sway! &c.

With breeches and jerkin of good country gray,
And live without working, now my strength doth decay;
With a hogshead of Sherry, for to drink when I please;
With friends to be merry, and to live at my ease;
    May I govern my Passion with an absolute sway! &c.

Without molestation, may I spend my last days
In sweet recreation; and sound forth the praise
Of all those that are true to the King and his laws!
Since it be their due, they shall have my applause!
    May I govern my Passion with an absolute sway! &c.

[When the days are grown short, and it freezes and snows;
May I have a coal fire as high as my nose!
A fire which, once stirred up with a prong,
Will keep the room temperate all the night long.
    May I govern my Passion with an absolute sway! &c.]

With courage undaunted, may I face my last day;
And, when I am dead, may the better sort say,
'In the morning, when sober; in the evening, when mellow:
He is gone, and has left not behind him his fellow!
    For he governed his Passion with an absolute sway!
    And grew wiser and better, as his strength wore away,
    Without gout or stone, by a gentle decay.'

ON THE DEATH OF A LADY'S DOG.

THOU, happy creature! art secure
From all the troubles we endure.
Despair, Ambition, Jealousy,
Lost Friends, nor Love, disquiet thee!
A sullen prudence drew thee hence,
From noise, fraud, and impertinence.
Though life essayed the surest wile,
Gilding itself with LAURA's smile;
How didst thou scorn life's meaner charms!
Thou, who couldst break through LAURA's arms!
Poor Cynic! Still, methinks, I hear
Thy awful murmurs in my ear;
As when on LAURA's lap you lay,
Chiding the worthless crowd away.
How fondly human Passions turn!
What then we envied, now we mourn!

ON A YOUNG LADY,

WHO SANG FINELY; AND WAS AFRAID OF A COLD.

WINTER! thy cruelty extend
   Till fatal tempests swell the sea!
   In vain, let sinking Pilots pray!
Beneath this yoke, let Nature bend!
   Let piercing frost and lasting snow,
   Through woods and fields destruction sow!

Yet we, unmoved, will sit and smile
   While you, these lesser ills create.
   These we can bear; but, gentle Fate,
And thou blest Genius of our Isle,
   From Winter's rage defend her Voice!
   At which the list'ning Gods rejoice.

May that celestial sound, each day,
   With ecstasy transport our souls!
   While all our Passions it controls;
And kindly drives our cares away.
   Let no ungentle cold destroy
   All taste we have of heavenly joy!

———.

UPON SOLITUDE.

HAIL, sacred Solitude! From this calm bay,
  I view the World's tempestuous sea!
    And, with wise pride, despise
    All those senseless vanities;
With pity moved for others, cast away
On rocks of Hopes and Fears. I see them tost
On rocks of Folly; and of Vice, I see them lost!
Some the prevailing malice of the Great
    (Unhappy men!), or adverse fate,
Sunk deep into the gulfs of an afflicted state:
But more, far more! a numberless prodigious Train,
  Whilst Virtue courts them (but, alas, in vain!),
    Fly from her kind embracing arms,
Deaf to her fondest call! blind to her greatest charms!
  And sunk in pleasures and in brutish ease,
They, in their shipwrecked state, themselves obdurate
    please.

  Hail, sacred Solitude! Soul of my soul,
    It is by thee I truly live!
Thou dost me better life and nobler vigour give!
  Dost each unruly appetite control!
  Thy constant quiet fills my peaceful breast
  With unmixed joy, uninterrupted rest!
    132

Presuming Love does ne'er invade)
This private solitary shade! |
And, with fantastic wounds by Beauty made,)
The joy has no alloy of jealousy, hope, and fear,
The solid comforts of this happy sphere.
Yet I, exalted Love admire!
Friendship, abhorring sordid gain;
And purified from Lust's dishonest stain!
Nor is it for my solitude unfit;
For I am with my friend alone,
As if we were but one!
'Tis the polluted Love that multiplies;
But Friendship does two souls in one comprise!

Here, in a full and constant tide, doth flow
All blessings Man can hope to know!
Here, in a deep recess of thought, we find
Pleasures which entertain, and which exalt, the mind!
Pleasures which do from friendship, and from know-
ledge, rise;
Which make us happy, as they make us wise!
Here, may I always, on this downy grass,
Unknown, unseen, my easy minutes pass!
Till, with a gentle force, victorious Death
My solitude invade;
And, stopping for a while my breath,
With ease convey me to a better shade.

FAIR Virtue! should I follow thee,
  I should be naked and alone!
For thou art not in company;
  And scarce art to be found in one!

Thy rules are too severe and cold
  To be embraced by vig'rous Youth!
And Fraud and Avarice arm the old
  Against thy justice and thy truth!

He who, by light of Reason led,
  Instructs himself in thy rough School,
Shall, all his lifetime, beg his bread;
  And, when he dies, be thought a fool!

Though, in himself, he 's satisfied
  With a calm mind and cheerful heart;
The World will call his virtue, Pride!
  His holy life, Design and Art!

The reign of Vice is absolute!
  While good men vainly strive to rise;
They may declaim! they may dispute!
  But shall continue poor and wise.

Honours and Wealth were made by Fate
  To wait on fawning Impudence!
To give insipid Coxcombs weight;
  And to supply the want of sense!

Mighty POMPEY, whose great soul
  Aimed at the liberty of Rome,
In vain did CAESAR's Arms control;
  And at Pharsalia was o'ercome.

His virtue, constant in distress,
  In PTOLEMY no pity bred;
Who, barely guided by success,
  Secured his peace, with his friend's head.

BRUTUS, whom the Gods ordained
  To do what POMPEY would have done,
The gen'rous motion entertained;
  And stabbed the tyrant on his throne!

This Godlike BRUTUS, whose delight
  Was Virtue, which he had adored,
Haunted by spectres overnight,
  Fell, the next day, on his own sword.

If, when his hope of vict'ry lost,
  This noble Roman could exclaim,
'O, Virtue! whom I courted most,
  I find she 's but an empty name!';

In a degen'rate Age like this,
  We, with more reason, may conclude,
'That Fortune will attend on Vice;
  And Misery, on those who dare be good!'

SYLVIA.

THE Nymph that undoes me, is fair and unkind;
No less than a wonder by Nature designed.
She 's the grief of my heart, the joy of my eye;
And the cause of a flame that never can die!

Her mouth, from whence wit still obligingly flows,
Has the beautiful blush, and the smell, of the rose.
Love and Destiny both attend on her will;
She wounds with a look; with a frown, she can kill!

The desperate Lover can hope no redress;
Where Beauty and Rigour are both in excess!
In SYLVIA they meet; so unhappy am I!
Who sees her, must love; and who loves her, must die!

A DRINKING SONG.

THE pleasures of Love and the joys of good Wine,
To perfect our happiness, wisely we join!
    We, to Beauty, all day,
    Give the sovereign sway;
And her favourite Nymphs devoutly obey!
At the Plays, we are constantly making our Court;
And when they are ended, we follow the sport
    To the Mall, and the Park;
    Where we love till 'tis dark!
    Then, sparkling Champagne

Puts an end to their reign.
It quickly recovers
Poor languishing Lovers!
Makes us frolic and gay; and drowns all our sorrow!
But, alas! we relapse again on the morrow!
Let every man stand
With his Glass in his hand;
And briskly discharge, at the word of command!
Here 's a Health to all those,
Whom, to-night, we depose!
Wine and Beauty, by turns, great souls should inspire!
Present all together! and now, boys, give fire!

VOITURE'S URANIA.

Hopeless, I languish out my days;
  Struck with Urania's conqu'ring eyes!
The wretch, at whom she darts these rays,
  Must feel the wound until he dies!

Though endless be her cruelty;
  Calling her beauties to my mind,
I bow beneath her tyranny,
  And dare not murmur 'She 's unkind!'

Reason, this tameness does upbraid;
  Proff'ring to arm in my defence:
But when I call her to my aid,
  She 's more a traitor than my Sense!

No sooner I the war declare,
    But, straight, her succour she denies;
And joining forces with the Fair,
    Confirms the conquest of her eyes.

———

CEASE, anxious World! your fruitless pain
    To grasp forbidden store!
Your studied labours shall prove vain,
    Your alchemy, unblest;
Whilst seeds of far more precious ore
    Are ripened in my breast.

My breast, the Forge of happier Love,
    Where my LUCINDA lives,
And the rich stock does so improve,
    As she her art employs,
That ev'ry smile and touch she gives,
    Turn all to Golden Joys!

Since then, we can such treasures raise;
    Let 's no expense refuse!
In Love, let 's lay out all our days!
    How can we e'er be poor;
When ev'ry blessing that we use
    Begets a thousand more?

TO A LADY

ASKING HIM, HOW LONG HE WOULD LOVE HER?

IT is not, CELIA, in our power
  To say, how long our love will last!
It may be, we, within this hour,
  May lose those joys we now do taste!
    The Blessed, that immortal be,
    From change in love are only free!

Then, since we mortal Lovers are,
  Ask not, How long our love will last?
But while it does; let us take care
  Each minute be with pleasure past!
    Were it not madness, to deny
    To live; because w' are sure to die!

———

FAIR IRIS! all our time is spent
  In trifling, whilst we dally.
The Lovers who're indifferent,
  Commit the grossest folly!
Ah! stint not then the flowing pleasure
To such a wretched scanty measure!
Since boundless Passion, boundless joys will prove;
Excess can only justify our love!

Excess, in other things so bad,
   In Love 's the justest measure!
No other reason 's to be had
   In that seraphic pleasure!
From growing love, bright Nymphs! your faces
Receive ten thousand sweeter graces!
My IRIS! then, that you may be divine,
Let your soft flame spread, night and day, like mine!

———

TELL me no more, I am deceived!
   While SYLVIA seems so kind,
And takes such care to be believed;
   The cheat I fear to find!
To flatter me, should Falsehood lie
   Concealed in her soft youth;
A thousand times, I'd rather die
   Than see the unhappy truth!

My love, all malice shall outbrave!
   Let Fops in libels rail!
If she, the appearances will save;
   No scandal can prevail!
She makes me think, I have her heart!
   How much for that, is due?
Though she but act the tender part,
   The joy she gives is true!

TO A VERY YOUNG LADY.

SWEETEST Bud of Beauty! may
No untimely frost decay
Th' early glories, which we trace
Blooming in thy matchless face!
But kindly opening, like the rose,
Fresh beauties, every day disclose;
Such as by Nature are not shown
In all the blossoms she has blown:
And then, what conquest shall you make;
Who hearts already daily take!

Scorched, in the Morning, with thy beams;
How shall we bear those sad extremes
Which must attend thy threat'ning eyes,
When thou shalt to thy Noon arise?

THE FORSAKEN MISTRESS.

PHILLIS.　　*TELL me, gentle STREPHON, why*
You, from my embraces fly?
Does my love, thy love destroy?
Tell me! I will yet be coy!

Stay, O, stay! and I will feign
(Though I break my heart!) disdain:
But, lest I too unkind appear;
For ev'ry frown, I'll shed a tear!

141

And if, in vain, I court thy love;
Let mine, at least, thy pity move!
Ah! while I scorn; vouchsafe to woo!
Methinks, you may dissemble too!

STREPHON.   AH! PHILLIS! that you would contrive
A way to keep my love alive!
But all your other charms must fail,
When kindness ceases to prevail!
Alas! no less than you, I grieve
My dying flame has no reprieve!
For I can never hope to find
(Should all the Nymphs I court, be kind!)
One Beauty able to renew
Those pleasures I enjoyed in you,
When Love and Youth did both conspire
To fill our breasts and veins with fire!

'Tis true, some other Nymph may gain
That heart, which merits your disdain!
But Second Love has still allay!
The joys grow agèd, and decay!
Then, blame me not for losing more
Than Love and Beauty can restore!
And let this truth thy comfort prove!
I would, but can no longer, love!

———

GATTY'S SONG.

To little, or no, purpose, I spent many days
In ranging the Park, th' Exchange, and th' Plays;
For ne'er, in my rambles, till now did I prove
So lucky to meet with the man I could love!
O, how I am pleased, when I think on this man,
That I find I must love; let me do what I can! . . .

———

YE happy Swains, whose hearts are free
   From Love's imperial chain,
Take warning, and be taught by me,
   T' avoid th' enchanting pain!
Fatal the wolves, to trembling flocks!
   Fierce winds, to blossoms prove!
To careless seamen, hidden rocks!
   To human quiet, Love!

Fly the Fair Sex! if bliss you prize.
   The snake 's beneath the flower!
Who ever gazed on beauteous eyes,
   That tasted quiet more?
How faithless is the Lovers' joy!
   How constant is their care!
The Kind, with falsehood do destroy;
   The Cruel, with despair!

———

BEAUTY NO ARMOUR AGAINST LOVE.

LADIES! though to your conqu'ring eyes
LOVE owes his chiefest victories;
And borrows those bright Arms from you,
With which he does the World subdue:
Yet you yourselves are not above
The empire, nor the griefs, of LOVE!

Then, wrack not Lovers with disdain;
Lest LOVE, on you revenge their pain!
You are not free, because y'are fair;
The Boy did not his mother spare!
Beauty 's but an offensive dart!
It is no armour for the heart.

THE FEMALE WITS.

BY A LADY OF QUALITY.

MEN, with much toil, and time, and pain,
   At length, at fame arrive;
While we, a nearer way obtain
   The palms for which they strive.

We scorn to climb, by Reason's rules,
   To the loud name of Wit;
And count them silly modest fools,
   Who to that test submit.

Our sparkling way, a method knows,
   More airy and refined;
And should dull Reason interpose,
   Our lofty flight 'twould bind!

Then, let us on! and still believe!
   A good bold faith will do!
If we ourselves can well deceive;
   The World will follow too!

What matter! though the witty few,
   Our emptiness do find;
They, for their int'rest, will be true!
   'Cause we are brisk and kind.

CATCHES.

Upon Christ Church bells in Oxford.

O, THE bonny Christ Church bells!
One! Two! Three! Four! Five! Six!
They sound so woundy great, so wondrous sweet;
And they troll so merrily! merrily!
    O, the First and Second Bell,
That, every day, at four and ten, cry
    'Come, come, come, come to Prayers!'
And the Verger troops before the Dean.
Tinkle! Tinkle! Ting! goes the small Bell
    At nine, to call the Beerers home:
But the Devil a man will leave his can,
    Till he hears the mighty Tom!

———

On Tobacco.

To be sung by four men smoking pipes.

GOOD, good indeed! The herb 's good weed!
Fill thy pipe, WILL; and I prithee, SAM, fill!
For, sure, we may smoke; and yet sing still!
146

What say the Learned? *Vita fumus.*

'Tis what {you and I, and he and I, you, and he, and I,} and all of us, *sumus.*

But, then, to the Learned say we again,
'If Life 's a Smoke, as they maintain,
If Life 's a Vapour; without doubt,
When a man does die, they should not cry,
" That his Glass is run: but his Pipe is out!"'

But whether we smoke, or whether we sing;
Let 's be loyal, and remember the King!
Let him live! and let his foes vanish
Thus! thus! thus! like a pipe of Spanish!

———

Upon Nothing.

Sing merrily now, my Lads! Here 's a Catch
  That was never meant, you!
But came by the Wheel of Fortune,
  Without any design, or intent, you.

It happened that, once, the Author,
  His head was exceeding hot,
A Catch he resolved he would make;
  And he couldn't tell of what!

He thought of the Smoke the weed affords;
   And it vanished all away!
He thought of fine Ladies and their fine Lords;
   And yet he found nothing to say!

He thought of a Thousand Pounds;
   But it wouldn't turn to account!
He thought of the Pot, and he thought of the Plot;
   But nothing would come on 't!

At last, he resolved, though nothing would do;
   That nothing should put him by, Sir!
But nothing to purpose, of Nothing he'd write;
   And nobody should be the wiser!

'Tis nothing to you, if he would do so!
   And if nothing is in 't, you find;
Then thank him for nothing! and that will be more
   Than ever he designed.

———————

FAIR ARCHABELLA! to thy eyes,
That flame like blushes in the skies,
Each noble heart doth sacrifice!

Yet be not cruel! since you may,
Whene'er you please to save, or slay;
Or with a frown benight the day!

I do not wish that you should rest
In any unknown highway breast,
The lodging of each common guest:

But I present a bleeding heart,
Wounded by love, not pricked by art,
That never knew a former smart!

Be pleased to smile; and then I live!
But if a frown, a death you give;
For which it were a sin to grieve.

Yet, if it be decreed I fall,
Grant me but one boon! One boon is all!
That you would me, your Martyr call.

Ye Virgin Powers! defend my heart
    From am'rous looks and smiles!
From saucy Love, or nicer Art
    Which most our sex beguiles!

From sighs and vows, from awful fears,
    That do to pity move!
From speaking silence; and from tears,
    Those springs that water Love!

But if, through Passion, I grow blind;
    Let Honour be my guide!
And where frail nature seems inclined,
    There place a guard of Pride!

A heart whose flames are seen, though pure,
    Needs every virtue's aid!
And those who think themselves secure;
    The soonest are betrayed!

STREPHON has Fashion, Wit, and Youth;
　With all things else that please;
He nothing wants but Love and Truth,
　To ruin me with ease!

But he is flint! and bears the art
　To kindle strong desire;
His power inflames another's heart,
　Yet he ne'er feels the fire!

Alas! it does my soul perplex,
　When I his charms recall,
To think he should despise the Sex;
　Or, what 's worse, love them all!

My wearied heart, like NOAH's dove,
　In vain may seek for rest!
Finding no hope to fix my love,
　Returns into my breast!

SONGS SUNG BY EIRENE.

TELL me, ye softer Powers above!
  Tell me, What unfledged thing
Begins within my breast to move,
  And try its tender wing?

Tell me, Why this unusual heat
  Thus creeps about my heart!
And why that heart indulges it,
  And fondly takes its part?

What Godhead could PHILANDER melt
  To such a flood of sighs;
That, gliding with the tide unfelt,
  He might my soul surprise?

Perfidious Music took my ear,
  And bent it to his Song!
Music, my friend, my darling care,
  Betrayed me on his tongue!

———

Anonymous.

O, MY PHILANDER! ope your breast!
 I can no longer keep my heart!
Why do you call it from its nest,
 With such a soft resistless art?

It sighs, and looks itself away!
 Dissolving with each word I speak.
O, take it! take it! If you stay,
 You will have nothing left to take!

There will be no injustice done,
 Though you have fired its native house,
If you will lodge it in your own;
 Where it can only find repose.

And there I'll rest, secure from harm,
 Let angry winds roar as they will!
That tongue can ev'ry tempest charm!
 Those eyes, the blackest cloud dispel!

LONG AND SHORT LIFE.

CIRCLES are praised, not that abound
In largeness; but th' exactly round:
So Life we praise, that does excel
Not in much time, but acting well!

OF THE LAST VERSES IN THE BOOK.

WHEN we, for age, could neither read nor write;
The Subject made us able to indite!
The Soul, with nobler resolutions decked,
The body stooping, does herself erect!
No mortal parts are requisite to raise
Her, that, unbodied, can her Maker praise!

The seas are quiet, when the winds give o'er;
So calm are we, when Passions are no more!
For then, we know how vain it was to boast
Of fleeting things, so certain to be lost!
Clouds of affection, from our younger eyes
Conceal that emptiness, which Age descries.

The Soul's dark cottage, battered and decayed,
Lets in new light, through chinks that Time hath
Stronger, by weakness; wiser, men become [made!
As they draw near to their eternal home.
Leaving the Old, both Worlds at once they view,
That stand upon the threshold of the New.

Miratur limen Olympi. VIRGIL.

O, LOVE! that stronger art than Wine,
Pleasing delusion! witchery divine!
Want, to be prized above all Wealth!
Disease, that has more joys than Health!
Though we blaspheme thee, in our pain;
And of thy tyranny complain:
We all are bettered by thy reign!

What Reason never can bestow,
We to this useful Passion owe!
LOVE wakes the dull from sluggish ease;
And learns the Clown the art to please!
Humbles the vain, kindles the cold!
Makes misers free, and cowards bold!
'Tis he reforms the sot from drink;
And teaches airy Fops to think!

When full brute appetite is fed;
And choked the glutton lies, and dead:
Thou, new spirits dost dispense,
And 'fines the gross delights of Sense.
Virtue's unconquerable aid!
That, against Nature can persuade;
And make the roving mind retire
Within the bounds of just desire.
Cheerer of Age! Youth's kind unrest!
And half the Heaven of the Blest!

CUPID IN CHAINS.

I SAW, last night, a pretty sight!
   CUPID a weeping lay,
Until his little eyes so bright
   Had wept themselves away!

I stepped unto him. 'Boy!' said I,
   'What causes all this mourning?'
He wiped his face, and then replied,
   'AMYNTAS still is scorning!

'The Youth defies my power, and cries,
   "I am a foolish Boy!"
He says, "I'm blind, and have no eyes;
   My deity 's a toy!"

'And as, last night, I sleeping lay,
   Down by yon crystal spring,
He came, and stole my bow away;
   And pinioned too my wing!'

'Alas!' cried I, ''twas then *thy* bow,
   Wherewith he wounded me!
I might have thought, that such a blow
   Could come from none but thee!

'Of late, he has, with too much art,
    Usurped divinity!
And plays the tyrant on that heart
    That yields itself to thee.

'Yet this, I'll for AMYNTAS plead.
    Since you must have your due,
Though he could all the captives lead;
    They're slaves to none but you!

'But thou, at last, revenged mayst be
    Upon th' ambitious Swain!
I'll set thy wings at liberty,
    And thou shalt fly again!

'And for this service on my part,
    I only beg of thee,
That thou wouldst wound AMYNTAS' heart;
    And make him die for me!'

He promised fair, while I untied
    His wings: but, waking, found
'Twas nothing but a dream! Alas!
    My heart had got the wound!

———

THE RETURN.

[*To J. Hoyle, Esquire.*]

AMYNTAS! whilst you
Have an art to subdue,
And can conquer a heart with a look, or a smile;
You pitiless grow,
And no faith will allow!
Tis the glory you seek, when you rifle the spoil.

Your soft warring eyes,
When prepared for the prize,
Can laugh at the aids of my feeble disdain!
You can humble the foe,
And soon make her know,   [vain!
Though she arms her with pride, her efforts are but

But, Shepherd! beware!
Though a victor you are,
A tyrant was never secure on his throne!
Whilst proudly you aim
New conquests to gain;
Some hard-hearted Nymph may return you your own!

Love, in fantastic Triumph sat,
　Whilst bleeding hearts around him flowed:
For whom fresh pains he did create;
　And strange tyrannic power he showed.
From thy bright eyes he took his fires;
　Which round about in sport he hurled:
But 'twas from mine he took desires,
　Enough t' undo the am'rous World.

From me, he took his sighs and tears!
　From thee, his pride and cruelty!
From me, his languishments and fears;
　And every killing dart from thee!
Thus, thou and I, the God have armed;
　And set him up a deity!
But my poor heart alone is harmed;
　Whilst thine the victor is, and free!

———

THE INVITATION.

DAMON! I cannot blame your will!
'Twas chance, and not design, did kill!
For whilst you did prepare your charms,
  On purpose SYLVIA to subdue;
  I met the arrows, as they flew,
    And saved her from their harms.

Alas! she cannot make returns!
Who for a Swain already burns;
A Shepherd whom she does caress
  With all the softest marks of love:
  And 'tis, in vain, thou seek'st to move
    The cruel Shepherdess!

Content thee, with this victory!
Think me as fair and young as she!
I'll make thee garlands all the day;
  And in the groves we'll sit and sing!
  I'll crown thee with the pride o' th' Spring!
    When thou art Lord of May.

———

CEASE, cease, AMINTA! to complain!
 Thy languishment give o'er!
Why shouldst thou sigh, because the Swain
 Another does adore?
Those charms, fond Maid! that vanquished thee,
 Have many a conquest won!
And, sure, he could not cruel be,
 And leave them all undone!

The Youth, a noble temper bears,
 Soft and compassionate;
And thou canst only blame thy stars,
 That made thee love too late!
Yet had their influence all been kind;
 They had not crossed my fate!
The tend'rest hours must have an end;
 And Passion has its date!

The softest love grows cold and shy!
 The face, so late adored,
Now, unregarded, passes by;
 Or grows, at last, abhorred!
All things in Nature fickle prove;
 See, how they glide away!
Think, so in time thy hopeless love
 Will die, as flowers decay!

THE Gods are not more blest than he
Who, fixing his glad eyes on thee,
With thy bright rays his senses cheers;
And drinks, with ever-thirsty ears,
The charming music of thy tongue;
Does ever hear, and ever long!
That sees, with more than human grace,
Sweet smiles adorn thy angel face.

But when with kinder beams you shine,
And so appear much more divine;
My feeble sense and dazzled sight
No more support the glorious light,
And the fierce torrent of delight!
O, then, I feel my life decay!
My ravished soul then flies away!
Then, faintness does my limbs surprise;
And darkness swims before my eyes!

Then, my tongue fails; and from my brow,
The liquid drops in silence flow!
Then, wand'ring fires run through my blood!
Then, cold binds up the languid flood!
All pale and breathless then I lie!
I sigh! I tremble! and I die!

THE DESPAIR.

IN a sad unfrequented cypress grove,
With all the symptoms of neglected love,
 The fair URANIA lay
By a clear murmuring river's side;
Her tears increasing the swift tide.
 With gales of sighs, I heard her say,
'Some pitying Power! O, ease my smart;
Or break at once my wretched heart!'

Then, still as Death the Virgin sat,
 Lost in a maze of thought,
But rousing with a sudden start,
 Ushered by a sad groan,
In charming sounds, her lute she taught
 Her killing grief to moan.

Thus sang the Fair! 'Ye Gods, it cannot be!
AMINTAS is not, can't be, false to me!
AMINTAS, he who, on my panting breast,
 So oft has leaned his sighing head!
 And things so soft, so tender, said,
As robbed me of my heart; and robbed me of my
  rest!
 So oft he vowed, that I believed!
For with that tongue, the World might be deceived!

'He wooed, he wan, with such an art,
 To LOVE himself unknown!
Should he the fatal way impart,
 All Maids were, sure, undone!
Yet, Heaven! to this poor, perjured Swain,
 Grant all the blessings in your powers!
Health to his flocks! and may no stain
Of falsehood blot his much-loved name,
 That name URANIA so adores!
Give him a fairer——' Nymph almost she said;
  But stopping, cried,
'Give him a thousand, thousand, joys beside!'

NEPTUNE'S RAGING FURY,

OR

THE GALLANT SEAMEN'S SUFFERINGS.

BEING A RELATION OF THEIR PERILS AND DANGERS, AND OF THE EXTRAORDINARY HAZARDS THEY UNDERGO IN THEIR NOBLE ADVENTURES: TOGETHER WITH THEIR UNDAUNTED VALOUR AND RARE CONSTANCY IN ALL THEIR EXTREMITIES; AND THE MANNER OF THEIR REJOICING ON SHORE, AT THEIR RETURN HOME.

YOU, Gentlemen of England, that live at home at ease,
Full little do you think upon the dangers of the seas!
Give ear unto the Mariners; and they will plainly show
The cares and the fears when the stormy winds do blow.

All you that will be Seamen, must bear a valiant heart!
For, when you come upon the seas, you must not think to start!
Nor once to be faint-hearted, in hail, rain, or snow!
Nor to shrink, nor to shrink when the stormy winds do blow!

The bitter storms and tempests, poor Seamen must endure,
Both day and night, with many a fright! We seldom rest
    secure!
Our sleep, it is disturbed with visions strange to know;
And with dreams on the streams, when the stormy winds
    do blow.

Altered from Martin Parker.

In claps of roaring thunder, which darkness doth enforce,
We often find our ships to stray beyond our wonted course ;
Which causeth great distractions, and sinks our hearts full low.
'Tis in vain to complain when the stormy winds do blow.

Sometimes in NEPTUNE's bosom our ships are tossed in waves,
And all the men expecting the sea to be their graves ;
Then, up aloft she mounteth ; and down again so low !
Tis with waves, O, with waves, when the stormy winds do
    blow.

Then down again we fall to prayer, with all our might and
    thought !
When refuge all doth fail us, 'tis that must bear us out !
To GOD we call for succour ; for He it is, we know,
That must aid us, and save us, when the stormy winds do blow.

The Lawyer and the Usurer, that sit in gowns of fur,
In closets warm can take no harm ; abroad they need not stir !
When Winter fierce, with cold doth pierce, and beats with hail
    and snow ;
We are sure to endure when the stormy winds do blow !

We bring home costly merchandise, and jewels of great price,
To serve out English Gallantry with many a rare device.
To please the English Gallantry, our pains we freely show ;
For we toil, and we moil, when the stormy winds do blow.

We sometimes sail to the Indies, to fetch home spices rare;
Sometimes 'gain to France and Spain, for wines beyond
    compare !
Whilst Gallants are carousing in Taverns on a row,
Then we sweep o'er the deep, when the stormy winds do blow.

When tempests are blown over, and greatest fears are past,
Aye weather fair and temperate air, we straight lie down to
    rest :
But when the billows tumble, and waves do furious grow,
Then we rouse, up we rouse, when the stormy winds do blow.

If enemies oppose us, when England is at wars
With any foreign nations ; we fear not wounds and scars !
Our roaring guns shall teach them our valour for to know,
Whilst they reel, in the keel, when the stormy winds do blow.

We are no cowardly shrinkers ; but Englishmen true bred !
We'll play our parts like valiant hearts ; and never fly for
    dread !
We'll ply our business nimbly, where'er we come or go !
With our mates, to the Straits, when the stormy winds do blow.

Then, courage, all brave Mariners ! and never be dismayed,
Whilst we have bold Adventurers, we ne'er shall want a trade !
Our Merchants will employ us, to fetch them wealth, I know.
Then be bold, work for gold, when the stormy winds do blow !

When we return in safety, with wages for our pains,
The Tapster and Vintner will help to share our gains !
We will call for liquor roundly, and pay before we go ;
Then, we'll roar, on the shore, when the stormy winds do blow !

THE COMPLAINT.

I LOVE! I dote! I rave with pain!
　No quiet 's in my mind!
Though ne'er could be a happier Swain,
　Were SYLVIA less unkind!
For when, as long her chains I've worn,
　I ask relief from smart;
She only gives me looks of scorn!
　Alas! 'twill break my heart!

My rival 's rich in worldly store,
　May offer heaps of gold;
But, surely, I a Heaven adore,
　Too precious to be sold!
Can SYLVIA such a coxcomb prize
　For wealth, and not desert;
And my poor sighs and tears despise?
　Alas! 'twill break my heart!

When, like some panting hov'ring dove,
　I for my bliss contend,
And plead the cause of eager love;
　She coldly calls me 'Friend!'
Ah! SYLVIA! thus, in vain, you strive
　To act a Healer's part!
'Twill keep but ling'ring pain alive;
　Alas! and break my heart!

When on my lonely, pensive bed,
  I lay me down to rest;
In hope to calm my raging head,
  And cool my burning breast:
Her cruelty all ease denies!
  With some sad dream I start!
All drowned in tears I find my eyes;
  And breaking feel my heart!

Then rising, through the path I rove
  That leads me where she dwells;
Where, to the senseless waves, my love
  Its mournful story tells.
With sighs, I dew and kiss the door,
  Till morning bids depart;
Then vent ten thousand sighs and more!
  Alas, 'twill break my heart!

But, SYLVIA! when this conquest 's won,
  And I am dead and cold;
Renounce the cruel deed you've done!
  Nor glory, when 'tis told!
For ev'ry lovely gen'rous Maid
  Will take my injured part;
And curse thee, SYLVIA! I'm afraid,
  For breaking my poor heart!

———

THE INCHANTMENT.

I DID but look and love awhile;
   'Twas but for one half-hour!
Then, to resist I had no will;
   And now, I have no power!

To sigh, and wish, is all my ease!
   Sighs which do heat impart
Enough to melt the coldest ice;
   Yet cannot warm your heart!

O, would your pity give my heart
   One corner of your breast!
'Twould learn of yours, the winning art;
   And quickly steal the rest!

CATCHES

OF THE RESTORATION AGE.

THE LONDON CONSTABLE.

WHO comes there? Stand! Who comes there?
And come before the Constable!   [Stand!
We will know, What you are?
What makes you out so late?
Says the midnight Magistrate,
With a noddle full of ale,
In a wooden Chair of State.

Whence come you, Sir? and whither do you go?
You may be, Sir! a Jesuit, for aught I know!

You may as well, Sir! take me for a Mahometan!

He speaks Latin! Secure him! He's a dangerous man!

To tell you the truth, Sir! I am an honest Tory;
But here's a crown to drink, and there's an end
    of the story.

Good morrow, Sir! a civil man is always welcome!
Go, BARNABY BOUNCE, light the Gentleman home!

OLD CHIRON thus preached to his pupil ACHILLES:
' I'll tell you, young Gentleman! what the Fates' will is!

You, my boy! must go,
The Gods will have it so,
To the Siege of Troy:

Thence never to return to Greece again;
But, before those walls, to be slain.

Let not your noble courage be cast down:
But, all the while you lie before the town,
Drink, and drive care away! drink, and be merry!
You'll ne'er go the sooner to the Stygian ferry!'

———

'TIS Women make us love!
'Tis Love that makes us sad!
'Tis Sadness makes us drink;
And Drinking makes us mad!

———

A FAREWELL TO WIVES.

ONCE in our lives,
Let us drink to our Wives!
Though the number of them is but small.
God take the best;
And the Devil take the rest!
And so we shall be rid of them all.

CLORIS' CHARMS
DISSOLVED BY EUDORA.

NOT that thy fair hand
Should lead me from my deep despair;
Or thy love, CLORIS! end my care,
  And back my steps command!
But if, hereafter, thou retire
To quench with tears thy wand'ring fire,
  This Clue I'll leave behind!
  By which thou mayst untwine
    The Saddest Way
    To shun the day
  That ever Grief did find.

  First, take thy hapless way
Along the rocky northern shore,
Infamous for the matchless store
  Of wrecks within that Bay.
None o'er the cursèd beach e'er crossed,
Unless the robbed, the wrecked, or lost!
  Where on the strand lie spread
  The skulls of many dead!
    Their mingled bones,
    Among the stones,
  Thy wretched feet must tread!

The trees, along the coast,
Stretch forth to heaven their blasted arms;
As if they plained the North Wind's harms,
   And youthful verdure lost.
There stands a Grove of fatal yew;
Where sun ne'er pierced, nor wind e'er blew.
   In it, a Brook doth fleet.
   The noise must guide thy feet!
      For there 's no light;
      But all is night
And darkness that you meet.

Follow th' infernal wave
Until it spread into a Flood,
Poisoning the creatures of the wood.
   There, twice a day, a slave,
I know not for what impious thing,
Bears thence the liquor of that spring.
   It adds to the sad place
   To hear how, at each pace,
      He curses God!
      Himself! his load!
For such his forlorn case.

Next, make no noise, nor talk,
Until thou'rt past a narrow Glade;
Where light does only break the shade.
   'Tis a Murderer's Walk!
Observing this, thou need'st not fear!
He sleeps the day, or wakes elsewhere.

Though there 's no clock, or chime,
The hour he did his crime,
    His soul awakes!
    His conscience quakes;
And warns him, that 's the time!

Thy steps must next advance
Where Horror, Sin, and Spectres dwell;
Where the wood's shade seems turned to Hell.
    Witches here nightly dance;
And Sprites join with them, when they call.
The Murderer dares not view the Ball!
    For snakes and toads conspire
    To make them up a Quire;
        And for their light,
        And torches bright;
    The Fiends dance all on fire!

Press on, till thou descry,
Among the trees sad, ghastly, wan,
Thin as the shadow of a man,
    One that does ever cry,
'She is not; and She ne'er will be!
Despair, and Death, come swallow me!'
    Leave him, and keep thy way!
    No more thou now canst stray!
        Thy feet do stand
        In Sorrow's Land;
    Its Kingdom 's every way.

Here, gloomy light will show,
Reared like a Castle to the sky,
A horrid Cliff, there standing nigh,
　Shading a creek below.
In which recess, there lies a Cave,
Dreadful as Hell, still as the grave.
　Sea-monsters there abide
　The coming of the tide.
　　No noise is near,
　　To make them fear.
God Sleep might there reside!

But when the boisterous sea,
With roaring waves, resumes this cell,
You'd swear the thunders there did dwell;
　So loud he makes his plea!
　So tempests bellow under ground;
And echoes multiply the sound!
　This is the place I chose,
　Changeable, like my woes:
　　Now calmly sad,
　　Then raging mad;
As move my bitter throws.

Such dread besets this part;
That all the horrors thou hast past
Are but degrees to this, at last.
　The sight must break thy heart!
Here bats and owls, that hate the light,
Fly and enjoy eternal night!

176

Scales of serpents, fish-bones,
Th' adder's eye, and toad-stones,
   Are all the light
   Hath blest my sight;
Since first began my groans.

When thus I lost the sense
Of all the healthful World calls Bliss,
And held it joy, those joys to miss;
   When Beauty was offence:
Celestial strains did rend the air,
Shaking these Mansions of Despair.
   A Form divine and bright
   Struck Day through all that Night;
     As when Heaven's Queen,
     In Hell was seen,
With wonder and affright!

The monsters fled for fear!
The terrors of the cursèd Wood
Dismantled were; and where they stood,
   No longer did appear!
The gentle Power, which wrought this thing,
EUDORA was; who thus did sing.
   'Dissolved is CLORIS' Spell;
   From whence thy evils fell!
     Send her this Clue!
     'Tis there most due,
And thy fantastic Hell!'

IN PRAISE OF HUNTING.

Leaving the Town and Phillis.

Tell me no more of Venus, and her Boy;
His flaming darts, and her transporting joy!
With dreams of pleasure, they delude our mind;
Which pass more swiftly than the fleeting wind!
The bright, the chaste, Diana I'll adore!
She'll free my heart from Love's insulting power!
Through pleasing groves, and o'er the healthful plain,
She leads the innocent and happy Swain.
Then, farewell, guilty crowds, and empty noise!
I leave you, for more pure and lasting joys!
In stately woods, gilded with morning rays,
I'll teach the echoes, great Diana's praise!

On racks of Love distended,
  Here lies a faithful Swain!
Wishing his life were ended,
  Or some respite to his pain.

The plague of dubious fate
  Is an ill beyond enduring!
  If I am not worth your curing;
Kill me quickly with your hate!

But why should Wit and Beauty
    Be guilty of such crimes?
Sure, 'tis a woman's duty
    To be merciful sometimes!

With justice you may slay
    The ungrateful and aspiring!
    But the humble and admiring,
You should treat a nobler way!

———

ONLY tell her, that I love!
    Leave the rest to her, and Fate!
Some kind Planet from above
May, perhaps, her pity move!
    Lovers, on their stars must wait!
Only tell her, that I love!

Why, O, why should I despair?
    Mercy 's pictured in her eye!
If She once vouchsafe to hear;
Welcome, Hope! and farewell, Fear!
    She 's too good, to let me die!
Why, O, why should I despair?

———

TO THE RT. HON. JOHN, LORD CUTTS,

AT THE SIEGE OF NAMUR.

THE HARDY SOLDIER.

'O, why is Man so thoughtless grown?
　Why, guilty souls in haste to die?
Vent'ring the leap to the worlds unknown;
　Heedless, to Arms and blood they fly!

'Are lives but worth a soldier's pay?
　Why will ye join such wide extremes,
And stake immortal souls, in play
　At desperate Chance, and bloody games?

'Valour 's a nobler turn of thought,
　Whose pardoned guilt forbids her fears:
Calmly she meets the deadly shot,
　Secure of life beyond the stars:

'But Frenzy dares eternal Fate;
　　And, spurred with Honour's airy dreams,
Flies to attack th' Infernal Gate,
　　And force a passage to the flames.'

Thus, hov'ring o'er Namuria's plains,
　　Sang Heavenly Love, in GABRIEL's form.
Young THRASO left the moving strains;
　　And vowed to pray, before the storm.

Anon, the thundering trumpet calls.
　　'Vows are but wind!' the Hero cries,
Then swears by Heaven, and scales the walls:
　　Drops in the ditch, despairs, and dies.

———

THE RETIREMENT.

WELL! I have thought on 't ; and I find
   This busy World is non-sense all!
I here despair to please my mind,
   Her sweetest honey is so mixed with gall!
Come then, I'll try how 'tis to be alone!
Live to myself a while, and be my own!

I've tried! and bless the happy change!
   So happy, I could almost vow
Never from this retreat to range!
   For, sure, I ne'er can be so blest as now!
From all th' allays of bliss, I here am free;
I pity others, and none envy me!

Here, in this shady lonely grove,
   I sweetly think my hours away!
Neither with Business vexed, nor Love;
   Which in the World bear such tyrannic sway.
No tumults can, my close apartment find,
Calm as those seats above, which know no storm
     nor wind.

Let Plots and News embroil the State ;
    Pray, What 's that, to my books and me ?
Whatever be the Kingdom's fate ;
    Here, I am sure t' enjoy a Monarchy !
Lord of myself, accountable to none ;
Like the First Man in Paradise, alone !

While the ambitious vainly sue,
    And of the partial stars complain ;
I stand upon the shore, and view
    The mighty labours of the distant Main.
I'm flushed with silent joy ; and smile to see
The shafts of FORTUNE still drop short of me !

Th' uneasy pageantry of State,
    And all the plagues to Thought and Sense,
Are far removed ! I am placed by Fate,
    Out of the road of all Impertinence !
Thus, though my fleeting life runs swiftly on,
'Twill not be short ! because 'tis *all* my own.

THE ELEVATION.

Take wing, my Soul! and upwards bend thy flight
To thy originary Fields of Light!
   Here 's nothing, nothing here below
     That can deserve thy longer stay!
   A secret whisper bids thee go
     To purer air, and beams of native day!
Th' ambition of the tow'ring Lark outvie;
And, like him, sing, as thou dost upward fly!

How all things lessen, which my Soul before
Did, with the grovelling multitude, adore!
   Those pageant glories disappear,
     Which charm and dazzle Mortals' eyes!
   How do I, in this higher Sphere,
     How do I, mortals, with their joys, despise!
Pure, uncorrupted Element I breathe;
And pity their gross atmosphere beneath.

How vile, how sordid, here those trifles show
That please the tenants of that Ball below!
   But, ah! I've lost the little sight!
     The scene 's removed; and all I see
   Is one confused dark mass of night!
     What nothing was, now nothing seems to be!
How calm this region! How serene! How clear!
Sure, I some strains of heavenly music hear!

On! On! The task is easy now, and light!
No steams of Earth can here retard thy flight!
  ' Thou need'st not now, thy strokes renew!
    'Tis but to spread thy pinions wide,
    And thou, with ease, thy seat wilt view ;
      Drawn by the bent of the ethereal tide.'
'Tis so, I find. How sweetly on I move!
Not let by things below ; and helped by those above.

But see! To what new region am I come ?
I know it well! It is my native home!
  Here led I once a life divine ;
    Which did all good, no evil know!
  Ah! who would such sweet bliss resign
    For those vain shows which Fools admire below!
'Tis true! But don't of folly past complain ;
But joy to see these blest abodes again!

A good retrieve! But, lo, while thus I speak,
With piercing rays th' Eternal Day does break!
  The beauties of the Face divine
    Strike strongly on my feeble sight!
  With what bright glories does it shine!
    'Tis ONE IMMENSE AND EVER-FLOWING LIGHT!
Stop here, my Soul! Thou canst not bear more bliss ;
Nor can thy now raised palate ever relish less!

Jane Barker.

A PASTORAL DIALOGUE
BETWIXT TWO SHEPHERD BOYS.

FIRST BOY.   I WONDER, what ALEXIS ails?
    To sigh, and talk of darts!
Of charms, which o'er his soul prevails!
    Of flames and bleeding hearts!
I saw him yesterday, alone,
    Walk crossing of his arms,
And, cuckoo-like, was in a tone,
    'Ah! CŒLIA! ah! thy charms!'

SECOND BOY. Why, sure, thou'rt not so ignorant
    As thou wouldst seem to be!
Alas, the cause of his complaint
    Is all our destiny!
'Tis mighty LOVE's all-pow'rful bow,
    Which has ALEXIS hit.
A pow'rful shaft will hit us too,
    Ere we're aware of it!

FIRST BOY.   Love! Why, alas! I little thought
    There had been such a thing!
Only for rhyme it had been brought,
    When Shepherds use to sing.
I'm sure, whate'er they talk of Love;
    'Tis but conceit at most!
As fear, in the dark, our fancies move
    To think we see a ghost!

SECOND BOY. I know not; but, the other day,
    A wanton Girl there were,
Who took my stock-dove's eggs away;
    And blackbird's nest did tear.
Had it been thee, my dearest boy!
    Revenge I should have took;
But she, my anger did destroy,
    With the sweetness of her look.

FIRST BOY. So, t' other day, a wanton slut,
    As I slept on the ground,
A frog into my bosom put;
    My hands and feet she bound.
She hung my hook upon a tree;
    Then, laughing, bade me wake!
And though she thus abusèd me,
    Revenge I cannot take!

CHORUS. Let 's wish these overtures of State
    Don't fatal omens prove!
For those who lose the power to hate,
    Are soon made slaves to Love.

———

THE heart you left, when you took mine,
    Proves such a busy guest,
Unless I do all power resign,
    It will not let me rest!

187

It my whole family disturbs!
　　Turns all my Thoughts away!
My stoutest Resolutions curbs!
　　Makes Judgement too obey!

If Reason interpose her power,
　　Alas! so weak she is;
She 's checked with one small soft amour
　　And conquered with a kiss!

———

TO MY YOUNG LOVER

ON HIS VOW.

ALAS! why mad'st thou such a vow?
　　Which thou wilt never pay!
And promisedst that, from very now
　　Till everlasting day,
Thou mean'st to love, sigh, bleed, and die;
　　And languish out thy breath
In praise of my divinity,
　　To th' minute of thy death!

Sweet Youth! Thou know'st not, what it is
　　To be Love's votary!
Where thou must, for the smallest bliss,
　　Kneel, beg, and sigh, and cry!
Probationer, thou shouldst be first;
　　That thereby thou mayst try,
Whether thou canst endure the worst
　　Of Love's austerity! . . .

BEGONE, old Care! and I prithee, be gone from me!
For, i' faith! old Care! thee and I shall never agree!
'Tis long thou hast lived with me, and fain thou wouldst me
 kill;
But, i' faith! old Care! thou never shalt have thy will!

Second Version. 1734.

Begone, old Care! I prithee, be gone from me!
Begone, old Care! you and I shall never agree!
Long time you have been vexing me, and fain you would
 me kill;
But, i' faith! old Care! thou never shalt have thy will!

Too much Care will make a young man look grey;
And too much Care will turn an old man to clay.
Come, you shall dance; and I will sing! So merrily we will
 play;
For I hold it one of the wisest things to drive old Care away.

Third Version. 1798.

Begone, dull Care! I prithee, be gone from me!
Begone, dull Care! you and I shall never agree!
Long time thou hast been tarrying here, and fain thou
 wouldst me kill;
But, i' faith! dull Care! thou never shalt have thy will!

Too much Care will make a young man turn grey;
And too much Care will turn an old man to clay.
My wife shall dance; and I will sing! So merrily pass the day!
For I hold it one of the wisest things to drive dull Care away!

A NEW SONG.

'Ho! brother TEAGUE! dost hear de Decree!
   Lilli burlero bullen a la.
Dat we shall have a new Debittie.
   Lilli burlero bullen a la.
   Lero, lero, lero, lero, lilli burlero bullen a la.
   Lero, lero, lero, lero, lilli burlero bullen a la.

'Ho! by my shoul! it is a T[ALBO]T;
   Lilli burlero bullen a la.
And he will cut all de English t[hroa]t!
   Lilli burlero bullen a la, &c.

'Though, by my shoul! de Inglish do prat;
   Lilli burlero bullen a la.
De Law 's on dare side; and CHREIST knows what!
   Lilli burlero bullen a la, &c.

'But if *Dispence* do come from the Pope;
   Lilli burlero bullen a la.
We'll hang *Magno Carto* and demselves in a rope!
   Lilli burlero bullen a la, &c.

'And the good T[ALBO]T is made a Lord;
   Lilli burlero bullen a la.
And he, with brave Lads, is coming aboard!
   Lilli burlero bullen a la, &c.

'Who! all in France have taken a swear,
  Lilli burlero bullen a la.
Dat day will have no Protestant h[ei]r!
  Lilli burlero bullen a la, &c.

'O, but why does he stay behind?
  Lilli burlero bullen a la.
Ho! by my shoul! 'tis a Protestant wind!
  Lilli burlero bullen a la, &c.

'Now, T[YRCONNE]L is come ashore;
  Lilli burlero bullen a la.
And we shall have Commissions gillore!
  Lilli burlero bullen a la, &c.

'And he dat will not go to M[a]ss,
  Lilli burlero bullen a la.
Shall turn out, and look like an ass!
  Lilli burlero bullen a la, &c.

'Now, now, de heretics all go down!
  Lilli burlero bullen a la.
By CHREIST! and St. PATRICK! the nation's our own!
  Lilli burlero bullen a la,' &c.
            [Printed in December 1688.]

THE SECOND PART.

THERE was an old prophecy, found in a bog,
  Lilli burlero bullen a la.
That Ireland be governed by an ass and a dog.
  Lilli burlero bullen a la, &c.

And now this old prophecy is come to pass.
  Lilli burlero bullen a la.
TALBOT is the dog; and TYRCONNEL's the ass!
  Lilli burlero bullen a la, &c.

'By CHRIEST! my dear MORISH, vat makes de sho' shad?
  Lilli burlero bullen a la.
'The heretics jeer us; and make me mad!
  Lilli burlero bullen a la, &c.

'Pox take me! dear TEAGUE! but I am in a rage!
  Lilli burlero bullen a la.
Poo! what impidence is in dish Age!
  Lilli burlero bullen a la, &c.

'Vat if Dush shou'd come, as dey hope,
  Lilli burlero bullen a la.
To up hang us, for all de *Dispence* of de Pope!
  Lilli burlero bullen a la, &c.

'Dey shay dat TYRCONNEL's a friend to de Mash:
  Lilli burlero bullen a la.
For which he 's a traitor, a pimp, and an ass!'
  Lilli burlero bullen a la, &c.

'Ara! plague tauke me! now I make a sware,
  Lilli burlero bullen a la.
I'd to Shent TYBURN will mauke a great prayer!
  Lilli burlero bullen a la, &c.

'O, I will pray to Shent PATRICK'S frock;
　Lilli burlero bullen a la.
Or to Loretto's sacred smock!
　Lilli burlero bullen a la, &c.

'Now a pox tauke me! What dost dow tink?
　Lilli burlero bullen a la.
De English, "Confusion to Popery!" drink.
　Lilli burlero bullen a la, &c.

'And, by my shoul! de Mash House pull down;
　Lilli burlero bullen a la.
While dey were swearing de Mayor of de town.'
　Lilli burlero bullen a la, &c.

'O, fait and be! I'll mauke de Decree,
　Lilli burlero bullen a la.
And sware, by de Chancellor's modesty!
　Lilli burlero bullen a la, &c.

'Dat I no longer in English will stay;
　Lilli burlero bullen a la.
For, be goad! dey will hang us out of the way!'
　Lilli burlero bullen a la, &c.

Printed in the year 1688
[*or rather* in January 1689].

TO MR. FINCH,

NOW EARL OF WINCHILSEA.

WHO, GOING ABROAD [I.E. OUT OF DOORS], HAD DESIRED
ARDELIA TO WRITE SOME VERSES UPON WHATEVER SUBJECT
SHE THOUGHT FIT, AGAINST HIS RETURN IN THE EVENING.

Written in the year 1689.

No sooner, FLAVIO! were you gone,
But, your injunction thought upon,
   ARDELIA took the pen;
Designing to perform the task,
Her FLAVIO did so kindly ask,
   Ere he returned again.

Unto Parnassus straight she sent,
And bid the messenger, that went
   Unto the Muses' Court,
Assure them, she their aid did need;
And begged they'd use their utmost speed,
   Because the time was short.

The hasty summons was allowed;
And, being well-bred, they rose and bowed,
   And said, ' They'd post away!
That well they did ARDELIA know;
And that no female's voice below,
   They sooner would obey!

'That many of that rhyming Train,
On like occasions, sought, in vain,
 Their industry t' excite:
But for ARDELIA, all they'd leave!'
Thus, flatt'ring, can the Muse deceive,
 And wheedle us to write!

'Yet, since there was such haste required,
To know the subject 'twas desired,
 On which they must infuse:
That they might temper words and rules;
And, with their counsel, carry tools,
 As Country Doctors use.'

Wherefore, to cut off all delays,
'Twas soon replied, 'A husband's praise!'
 Though in these looser Times,
ARDELIA gladly would rehearse
A husband's; who indulged her Verse,
 And now required her rhymes.

'A husband!' echoed all around!
And to Parnassus, sure, that sound
 Had never yet been sent!
Amazement in each face was read!
In haste, all the affrighted Sisters fled;
 And unto council went.

ERATO cried, 'Since GRISEL's days,
Since Troy town pleased, and *Chevy-Chase*,

No such design was known;
And 'twas their business to take care,
It reached not to the public ear,
  Or got about the Town:

'Nor came where evening Beaus were met
O'er *billets-doux* and chocolate;
  Lest it destroyed the House!
For, in that place, who could dispense
(That wore his clothes with common sense)
  With mention of a *Spouse*?'

'Twas put unto the vote at last;
And in the negative it past.
  None to her aid should move!
Yet, since ARDELIA was a friend,
Excuses 'twas agreed to send,
  Which plausible might prove.

'That Pegasus, of late, had been
So often rid through thick and thin,
  With neither fear nor wit;
In Panegyric, been so spurred;
He could not from the stall be stirred,
  Nor would endure the bit!

'MELPOMENE had given a bond,
By the new House alone to stand;
  And write of war and strife!
THALIA, she had taken fees
And stipends from the Patentees;
  And durst not, for her life!'

URANIA [1] only liked the choice:
Yet, not to thwart the Public Voice,
　　She, whispering, did impart,
'They need no foreign aid invoke,
No help to draw a moving stroke,
　　Who dictate from the heart!'

'Enough!' the pleased ARDELIA cried:
And, slighting ev'ry Muse beside,
　　Consulting now her breast,
Perceived that ev'ry tender thought,
Which from abroad she'd vainly sought,
　　Did there in silence rest;

And should unmoved that Post maintain
Till, in his quick return again,
　　Met in some neighb'ring grove
(Where vice, nor vanity appear),
Her FLAVIO them alone might hear
　　In all the sounds of love.

For since the World does so despise
HYMEN's endearments, and its ties;
　　They should mysterious be!
Till we that pleasure too possess
(Which makes their fancied happiness!),
　　Of stolen secrecy!

---

[1] URANIA is the Heavenly Muse; and is supposed to inspire thoughts
of virtue.

ON LOVE.

Love! thou'rt the best of human joys!
  Our chiefest happiness below!
All other pleasures are but toys!
Music, without thee, is but noise;
  And Beauty but an empty show!

Heaven, who knew what Man could move
  And raise his thoughts above the brute,
Said, 'Let him live! and let him love!'
'Tis this alone, that can his soul improve;
  Whate'er Philosophers dispute!

ANONYMOUS.

ALMIRA TO DAMON.

Why fear'st thou, Damon! I should stray,
Dazzled, or bribed, with glitt'ring clay?
The weak-eyed owls may lose their sight
At objects that are over-bright;
  Yet, sure, Almira's sounder sense
May baffle gold's sophistic eloquence!

198

This hurtful devil was confined,
By Heaven's indulgence to mankind,
Within the caverns of the earth ;
Where first he had his cursèd birth :
But we soon raised him from his nest ;
And now the tempting fiend disturbs our rest!

A just reward ! and may the slaves,
That dig the mines, dig their own graves !
Gold's jaundice and a woman's skin
Proclaim them both of different kin ;
And they deserve th' inherent woe,
That make alliances with such a foe !

ALMIRA bravely shall despise
Trifles, which sordid mortals prize!
Their earthly thoughts may downward roll
Prone to the Centre of their soul :
While mine, refined like heavenly fire,
Active and pure, still upwards will aspire !

Then if thou canst but equal me
In fervent love and constancy,
And canst find courage to despise
Those little worldly vanities ;
With thee, I'll spend my utmost breath !
With thee, I'll bravely scorn approaching Death !

Ah! blame me not, if no despair,
  (A Passion you inspire!) can end!
Nor think it strange, too charming Fair!
  If Love, like other flames, ascend!
If to approach a Saint with prayer
  Unworthy votaries pretend
Above all merit, Heaven and you
To the sincere are only due!

Long did respect awe my proud aim;
  And fear to offend, my madness cover.
Like you, it still reproved my flame;
  And in the Friend would hide the Lover.
But, by things that want a name,
  I the too bold truth discover!
My words, in vain, are in my power;
My looks betray me every hour!

Robert Gould.

Fair, and soft, and gay, and young,
All charm! She played, She danced, She sung!
There was no way to 'scape the dart!
No care could guard the Lover's heart!
'Ah! why,' cried I, and dropped a tear
(Adoring; yet despairing e'er
To have her to myself alone!),
'Was so much sweetness made for One?'

But, growing bolder, in her ear
I, in soft Numbers, told my care.
She heard; and raised me from her feet,
And seemed to glow with equal heat.
Like Heaven's, too mighty to express,
My joys could but be known by guess!
'Ah! fool!' said I, 'what have I done,
To wish her made for more than One!'

But long She had not been in view,
Before her eyes, their beams withdrew.
Ere I had reckoned half her charms,
She sank into another's arms!
But she that once could faithless be,
Will favour him no more than me!
He, too, will find he is undone;
And that She was not made for One!

FATAL CONSTANCY.

CLARA, charming without art,
  The wonder of the plain,
Wounded by LOVE's resistless dart,
Had over-fondly given her heart
  To a regardless Swain ;
    Who, though he well knew
    Her Passion was true,
Her truth and her beauty disdained ;
    While thus the fair Maid,
    By her folly betrayed,
To the rest of the Virgins complained.

'Take heed of Man ! and, while you may,
  Shun Love's deceitful snare !
For though, at first, it looks all gay ;
'Tis ten to one y' are made a prey
  To sorrow, pain, and care !
    But if you love first,
    Y' are certainly curst !
Despair will insult in your breast !
    The nature of Men
    Is to slight who love them ;
And love those that slight them, the best !

'Yet, let the conqu'ror know my mind!
  Ingrateful CELADON!
That he will never, never, find
One half so true, or half so kind;
  When I am dead and gone.'
    But, as she thus spoke,
    Her tender heart broke!
Death spares not the Fair; nor the Young!
    So swans, when they die,
    Make their own Elegy;
And breathe out their life in a Song.

———

NOT though I know he fondly lies
  Pressed in my rival's arms;
Nor though my friends, with tears, advise
  That I should shun his charms:
Nor one, nor t'other, frees my heart
  (Such arts he does display!);
Or can my longing eyes divert
  From gazing still that way.

Tell me, ye Powers that rule our fate!
  Why are we made so vain,
Most earnestly to wish for that
  We have least hope t' attain?
Or, if attained, is but, at best,
  A mine of rifled ore;
An empty cabinet, the breast,
  The jewel gone before! . . .

THE VICTORY IN HUNGARY.

HARK! how the Duke of LORRAINE comes!
   The brave victorious Soul of War!
With trumpets and with kettle-drums,
   Like thunder rolling from afar.

On the Left Wing, the conqu'ring Horse,
   The brave Bavarian Duke does lead.
These heroes, with united force,
   Fill all the Turkish host with dread.

Their bright caparisons behold!
   Rich habits, streamers, shining arms,
The glittering steel and burnished gold,
   The pomp of War with all its charms.

With solemn march, and fatal pace,
   They bravely on the foe press on!
The cannons roar, the shot takes place;
   Whilst smoke and dust obscure the sun.

The horses neigh, the soldiers shout;
  And now the furious Bodies join!
The slaughter rages all about;
  And men in groans their blood resign.

The weapons clash; the roaring drum,
  With clangour of the trumpets' sound;
The howls and yells of men o'ercome;
  And from the neighbouring hills rebound.

Now, now, the Infidels give place!
  Then, all in routs, they headlong fly!
Heroes, in dust, pursue the chase;
  While deaf'ning clamours rend the sky!

———

THE EXPOSTULATION.

'STILL wilt thou sigh! and still, in vain,
  A cold neglectful Nymph adore!
No longer fruitlessly complain;
  But to thyself, thyself restore!
In youth, thou caught'st this fond disease;
  And shouldst abandon it in age!
Some other Nymph as well may please!
  Absence, or business, disengage!'

'On tender hearts, the wounds of Love,
　　Like those imprinted on young trees,
Or kill at first, or else they prove
　　Larger b' insensible degrees!
Business I tried; she filled my mind!
　　On others' lips, my Dear I kissed!
But never solid joy could find,
　　Where I, my charming SILVIA missed.

'Long absence, like a Greenland night,
　　Made me but wish for sun the more;
And that inimitable light,
　　She, none but she, could e'er restore!'
'She never once regards thy fire;
　　Nor ever vents one sigh for thee!'
'I must the glorious sun admire;
　　Though he can never look on me!'

'Look well! You'll find she 's not so rare!
　　Much of her former beauty 's gone!'
'My love, her shadow, larger far
　　Is made by her declining sun!
What if her glories faded be;
　　My former wounds I must endure!
For should the Bow unbended be;
　　Yet that can never help the cure!'

WRITTEN ON

THE LEAVES OF A WHITE FAN,

BORROWED FROM MISS OSBORNE, AFTERWARDS HIS WIFE.

FLAVIA, the least and slightest toy
Can, with resistless art, employ!
This Fan, in meaner hands, would prove
An engine of small force in Love:
Yet she, with graceful Air and mien
(Not to be told, or safely seen!),
Directs its wanton motions so,
That it wounds more than CUPID's Bow!
Gives coolness to the matchless Dame;
To every other breast—a flame!

LIBERTY RETRIEVED.

How you oblige me, scornful Fair!
Such treatment makes me free as air!
Reason, at length, unseals my eyes;
And fond, mistaken Passion flies!

Late, I was full of wondrous flame;
Languished, and trembled at your name!
But Love, for want of Hope, expires;
And Rigour cools whom Beauty fires!

Your melting notes and radiant eyes
Made my soft heart a willing prize!
I blushed not to confess my chain,
And sang the triumphs of your reign!

But cruel VENUS has decreed,
That PHŒBUS, nor his race, succeed!
While to a MARS', or VULCAN's arms,
The Goddess straight resigns her charms!

Sir Fleetwood Sheppard.

No more I durst my pen advance;
Yet still presumed to court a glance!
Hoping you'd view my look of care,
And read my anguish in my Air.

You saw, 'tis true! but practised hate;
And seemed to mark me out for fate!
But, thanks to Heaven! a Briton born
Is not so mean to die by scorn!

At this, I mustered all the Man!
To just rebellion fearless ran!
Flung the proud tyrant from her throne;
And made the captived fort my own!

Let giddy Fops employ your arts!
Contrive to catch their worthless hearts!
Your charms can easily deceive!
Their vanity will soon believe!

But grant, perhaps, they own your sway;
They give but glories of a day!
Beauty 's, at best, a fading flower;
The spoil of ev'ry barbarous hour!

And when the rosy bloom of Youth
Leaves you, they leave their boasted truth!
To some new Charmer, fine things say;
And swear, 'They cannot court Decay!'

I've only lost enchanting pain:
But you, a faithful artless Swain,
That had been true; had you but known
To wear, as well as win, a crown!

'Tis done! I'll be no longer bound;
Nor let disdainful Woman wound!
O, would all men resolve like me;
No tyrant should a conqu'ror be!

I'd set up milder laws for Love;
And, by successful methods, prove,
The Face dependent on the Mind;
And no Nymph charming, but the kind!

Matthew Prior.

AN EPISTLE

TO SIR FLEETWOOD SHEPPARD.

Written *anno* 1689.

WHEN crowding folks, with strange ill faces,
Were making legs, and begging Places;
And some with Patents, some with merit,
Tired out my good Lord DORSET's spirit:
Sneaking, I stood among the crew,
Desiring much to speak with you.

I waited while the clock struck thrice,
And footman brought out fifty lies;
Till patience vexed, and legs grown weary,
I thought it was in vain to tarry.
But did opine, it might be better,
By Penny Post, to send a letter.

P 2

211

Now, if you miss of this Epistle,
I am balked again; and may go whistle!

My business, Sir! you'll quickly guess
Is to desire some little Place!
And fair pretensions I have for 't;
Much need, and very small desert!

Whene'er I writ to you, I wanted;
I always begged, you always granted!
Now as you took me up when little,
Gave me my learning and my victual;
Asked, for me, from my Lord things fitting,
Kind as I'd been your own begetting;
Confirm what formerly you've given!
Nor leave me now at six and sevens,
As SUNDERLAND has left MUN STEPHENS. . . .}

My uncle (rest his soul!), when living,
Might have contrived me ways of thriving!
Taught me, with cider to replenish
My vats, or ebbing tide of Rhenish!
So when, for Hock, I drew pricked White Wine,
Swear 't had the flavour, and was right wine!
Or sent me, with ten pounds, to Furni-
val's Inn, to some good rogue Attorney;

Where now, by forging deeds and cheating,
I'd found some handsome ways of getting!

All this, you made me quit! to follow
That sneaking whey-faced God APOLLO.
Sent me among a fiddling crew ⎞
Of folks, I'd never seen, nor knew; ⎬
CALLIOPE, and God knows Who! ⎠
    To add no more invectives to it;
You spoiled the Youth, to make the Poet! . . .

The sum of all I have to say ⎞
Is, That you put me in some way; ⎬
And your Petitioner shall pray—— ⎠

There 's one thing more, I had almost slipt;
But it may do as well in Postscript.
    My friend, CHARLES MONTAGU, 's preferred: ⎞
Nor would I have it long observed, ⎬
The *one Mouse* eats; while *t' other* 's starved! ⎠

——

Matthew Prior.

Whilst Beauty, Youth, and gay Delight
   In all thy looks and gestures shine;
Thou hast, my Dear! undoubted right
   To rule this destined heart of mine!
My reason bends to what your eyes ordain:
For I was born to love; and you, to reign!

But would you meanly then rely
   On power, you know I must obey?
It is but legal tyranny
   To do an ill, because you may!
Why must I, thee, as Atheists, Heaven adore?
Not see thy mercy; and but dread thy power!

Take care, my Dear! Youth flies apace!
   Time, equally with Love, is blind.
Soon must these glories of thy face,
   The fate of vulgar Beauties find!
The thousand Loves that arm thy potent eye,
Must drop their quivers, flag their wings, and die!

Then wilt thou sigh, when, in each frown,
　　One hateful wrinkle more appears,
And putting peevish humours on,
　　Seems but the sad effect of years.
E'en kindness, then, too weak a Charm will prove,
To reinflame the ashes of my love!

Forced compliments and formal bows
　　Will shew thee just above neglect.
The heat, with which thy Lover glows,
　　Will settle into cold respect.
A talking dull Platonic I shall turn!
Learn to be civil; when I cease to burn!

Then, shun that ill; and know, my Dear!
　　Kindness and Constancy will prove
The only pillars fit to bear
　　So vast a weight as that of Love!
If thou wouldst wish to make my flames endure;
Thine must be very fierce, and very pure!

Haste, CELIA! haste, while LOVE invites!
　　Obey the gentle Godhead's voice!
Fill ev'ry sense with soft delights;
　　And give thy soul a loose to joys!
Let millions of repeated blisses prove
That thou art Kindness all; and I, all Love!

Be mine; and only mine! Take care
    Thy words, thy looks, thy dreams, to guide
To me alone! nor come so far
    As liking any Youth beside!
What men e'er court thee; fly them! and believe
They're serpents all; and thou, the tempted EVE!

So will I court thy dearest truth,
    When beauty ceases to engage;
And, thinking on thy pleasing youth,
    I'll love thee on, in spite of age!
So time itself our transports shall improve;
And still we'll wake to joys, and live to love!

———

WHILST I am scorched with hot desire;
    In vain, cold friendship you return!
Your drops of pity on my fire,
    Alas! but make it fiercer burn!

Ah! would you have the flame supprest,
    That kills the heart it heats too fast;
Take half my Passion to your breast,
    The rest in mine shall ever last!

CORINNA, in the bloom of youth,
    Was coy to every Lover;
Regardless of the tenderest truth,
    No soft complaint could move her!

Mankind was hers! All, at her feet,
    Lay prostrate and adoring;
The Witty, Valiant, Rich, and Great,
    Alike in vain imploring.

But now grown old; she would repair
    Her loss of time and pleasure!
With willing eyes, and wanton Air,
    Inviting every gazer.

But Love 's a summer flower, that dies
    With the first weather's changing.
The Lover, like the swallow, flies
    From sun to sun, still ranging.

CLOE! Let this example move
    Your foolish heart to reason!
Youth is the proper time for Love;
    And Age is Virtue's season!

PREPARED to rail, resolved to part;
  When I approach the perjured Maid,
What is it awes my timorous heart?
  Why is my tongue afraid?

With the least glance a little kind
  (Such wondrous power have MYRA's charms!),
She drives my doubts! enslaves my mind!
  And all my rage disarms!

Forgetful of her broken vows,
  When gazing on that form divine,
Her injured vassal trembling bows;
  Nor dares the slave repine!

———

IN vain, a thousand slaves have tried
To overcome BELINDA's pride!
    Pity pleading,
    Love persuading.
When her icy heart is thawed;
Honour chides, and, straight, she 's awed!

    Foolish creature!
    Follow Nature!
  Waste not thus your prime!
    Youth 's a treasure,
    Love 's a pleasure;
  Both destroyed by Time!

———

AH! CELIA! why so fiercely bent
  To vex my tender heart?
To Gold and Title you relent;
  LOVE throws in vain his dart!

Let glittering fools in Courts be great!
  For pay, let armies move!
Beauty should have no other bait
  But tender vows and love!

If on those endless charms you lay
  The value that's their due;
Kings are themselves too poor to pay!
  A thousand worlds too few!

But if a Passion without vice,
  Without disguise, or art;
Ah! CELIA! if true love's your price;
  Behold it, in my heart!

SOME die with their eternal toil,
  Through pride, ambition, or for gain,
For fame, for empire, or for spoil;
  But I, by cruel Love am slain!

PROMETHEUS, on a mountain tied,
  Is doomed to suffer endless pain;
PHÆTON, by thunder died:
  But I, by cruel Love am slain!

ON A LOVER BEGINNING TO LOVE.

As free as wanton winds I lived,
   That unconcerned do play.
No broken faith, no fate I grieved;
   No fortune gave me joy!
A dull content crowned all my hours!
   My heart no sighs oppressed!
I called, in vain, on no deaf Powers,
   To ease a tortured breast!

The sighing Swains regardless pined,
   And strove in vain to please;
With pain, I civilly was kind:
   But could afford no ease.
Though Wit and Beauty did abound;
   The charm was wanting still,
That could inspire the tender wound,
   Or bend my careless will.

Till in my heart a kindling flame
   Your softer sighs had blown;
Which I, with striving, love, and shame,
   Too sensibly did own!
Whate'er the God before could plead,
   Whate'er the Youth's desert;
The feeble siege, in vain, was laid
   Against my stubborn heart.

At first, my sighs and blushes spoke,
  Just when your sighs would rise:
And when you gazed, I wished to look;
  But durst not meet your eyes!
I trembled, when my hand you pressed;
  Nor could my guilt control:
But love prevailed, and I confessed
  The secrets of my soul.

And, when upon the giving part,
  My present to avow,
By all the ways conformed my heart
  That Honour would allow.
Too mean was all that I could say!
  Too poorly understood!
I gave my soul the noblest way;
  My letters made it good!

———

  You, Lovers! love on!
  Lest the world be undone;
And Mankind be lost by degrees.
    For if all, from their Loves,
    Should go wander in groves;
There soon would be nothing but trees!

John Howe.

SHE.   *You say, 'Tis Love creates the pain,*
Of which so sadly you complain;
And yet would fain engage my heart
In that uneasy cruel part!
But how, alas! think you, that I
Can bear the wound, of which you die?

HE.   'Tis not my Passion makes my care;
But your indiff'rence gives despair!
The lusty sun begets no Spring,
Till gentle showers assistance bring;
So Love, that scorches and destroys,
Till Kindness aids, can cause no joys!

SHE.   *Love has a thousand ways to please;*
But more to rob us of our ease!
For wakeful nights, and careful days;
Some hours of pleasure he repays!
But absence soon, or jealous fears,
O'erflow the joys with floods of tears.

222

He.   By vain and senseless forms betrayed,
Harmless Love 's th' offender made:
While we no other pains endure,
Than those that we ourselves procure.
But one soft moment makes amends
For all the torment that attends.

Chorus.

Both. Let us love! Let us love, and to happiness haste!
Age and Wisdom come too fast!
Youth for loving was designed!

He alone. I'll be constant! You be kind!

She alone. You be constant! I'll be kind!

Both. Heaven can give no greater blessing
Than faithful love, and kind possessing!

WHENE'ER I sing, or on my flute I play;
  PHYLLIS! you pass whole hours to hear me!
But when I speak of love, you go away;
  And will not stay, alas! a moment near me!

O, hear a faithful Lover's gentle tale!
  Music, though sweet, 's not half so moving!
Lose not your bliss! Age will on Youth prevail;
  And then, you'll vainly wish to hear of loving!

———

DAMON! why will you die for love;
  Yet ne'er your flame discover?
Be wise! and soon that pain remove;
  Or tell the Nymph, you love her!
As in each other fierce disease,
  So in Love's cruel anguish:
He who wants sense to beg for ease,
  Deserves in pain to languish!

Women, like Fortune, love the bold!
  Like her, their minds they vary!
Perhaps, this day, though CELIA 's cold;
  With you, the next, she'll marry!
Be sure, be true! if she is kind;
  If cruel, then forget her!
With little pains, you soon will find
  A Nymph who'll use you better!

———

Boasting Fops, who court the Fair
　For the fame of being loved:
You who daily prating are
　Of the hearts your charms have moved;
Still be vain in talk and dress!
　But, while shadows you pursue,
Own that some, who boast it less,
　May be blessed as much as you!

Love and Birding are allied!
　Baits and nets alike they have!
The same arts in both are tried,
　The unwary to enslave.
If, in each, you'd happy prove;
　Without noise, still watch your prey!
For, in Birding and in Love,
　While we talk, it flies away!

Not yet bestowed! Melania? Why,
　Can nothing mortal please your eye?
Or can your piercing mind, with certain art,
　Break the dark closet of the breast,
Where unfledged perjuries in secret rest;
And, with one glance, read o'er the mischief-teeming
　　heart?

We know the utmost Wit can do;
　And we'd expect it all from you!
But virgin Innocence confines your skill.
　Alas! you only know the Good!
Nor more our happy Grandame understood,
Till the forbidden fruit had shewn the fatal Ill.

Yet, sure, our sex, though tainted, gives
　One soul where sacred Virtue lives!
When the devouring Deluge all around
　The first-born wretched world embraced;
One yet the great unerring Censor graced!
In the corrupted mass, One righteous man was found!

Choose then, Melania! choose the boy,
　Who must the glorious prize enjoy!
From your blessed union shall that Era rise,
　From whence the bettered World shall date
Her Golden Age; while Virtue, linked with Fate,
By your example, bids the bedlam World be wise.

226

So gold, while buried in the mine,
  May there with useless lustre shine:
But when drawn out, the royal stamp it wears,
  Through all the needy World it flies;
And all in bands of friendly commerce ties,
And gluts the miser's wish, and dries the beggar's
    tears.

———

You understand no tender vows
  Of fervent and eternal love!
That Lover will his labour lose,
Who does, with sighs and tears propose
    Your heart to move!
But if he talk of settling land,
A House in Town, and Coach maintained;
    You understand!

You understand no charms in Wit,
  In Shape, in Breeding, or in Air!
To any Fops you will submit,
The nauseous Clown, or fulsome Cit;
    If rich they are!
Who guineas can, may you, command!
Put gold; and then put in your hand!
    You understand!

AN ANNIVERSARY ODE TO

QUEEN MARY,

29TH OF APRIL, 1692.

Love's Goddess, sure, was blind, this day,
   Thus to adorn her greatest foe;
And Love's artillery betray
   To one, that would her realm o'erthrow.

Those Eyes, that form that lofty mien,
   Who could for Virtue's Camp design?
Defensive Arms should there be seen;
   No sharp, no pointed, weapons shine!

Sweetness of nature, and true wit,
   High power, with equal goodness joined,
In this fair Paradise are met—
   The joy and wonder of Mankind!

Long may she reign over this Isle,
   Loved and adored in foreign parts!
But, gentle PALLAS! shield the while
   From her bright charms our single hearts!

May her blest example chase
   Vice in troops out of the land!
Flying from her awful face,
   Like pale ghosts, when day 's at hand.

May her Hero bring home Peace,
   Won with honour in the Field!
And all home-bred factions cease!
   He, our Sword; and she, our Shield!

Many such Days may she behold,
   Like the glad sun, without decay!
May Time, that tears where he lays hold,
   Only salute her, in his way!

Late, late, may she to Heaven return!
   And Quires of Angels there rejoice
As much, as we below shall mourn
   Our short, but their eternal, choice!

SEE, HYMEN comes! How his torch blazes!
  Looser Loves, how dim they burn!
No pleasures equal chaste embraces;
  When we, love for love return.

When FORTUNE makes the Match, he rages;
  And forsakes th' unequal pair!
But when Love two hearts engages,
  The kind God is ever there!

Regard not, then, High Blood, nor Riches;
  You that would his blessings have!
Let untaught Love guide all your wishes!
  HYMEN should be CUPID's slave!

Young Virgins! that yet bear your Passions
  Coldly, as the flint her fire,
Offer to HYMEN your devotions!
  He will warm you with desire!

Young Men! no more neglect your duty
  To the God of nuptial vows!
Pay your long arrears to Beauty,
  As his chaster law allows!

———

PHILLIS KNOTTING.

'HEARS not, my PHILLIS! how the birds
  Their feathered mates salute!
They tell their Passion in their words;
  Must I alone be mute?'
        PHILLIS, without frown or smile,
        Sat and knotted all the while.

'The God of Love, in thy bright eyes,
  Does like a Tyrant reign:
But, in thy heart, a Child he lies,
  Without his dart, or flame.'
        PHILLIS, without frown or smile, &c.

'So many months, in silence past,
  And yet in raging love,
Might well deserve one word, at last,
  My Passion should approve!'
        PHILLIS, without frown or smile, &c.

'Must then your faithful Swain expire,
  And not one look obtain?
Which he, to soothe his fond desire,
  Might pleasingly explain.'
        PHILLIS, without frown or smile, &c.

———

231

Ah! Cloris! that I now could sit
    As unconcerned, as when
Your infant beauty could beget
    No pleasure, nor no pain!

When I the Dawn used to admire,
    And praised the coming Day;
I little thought the growing fire
    Must take my rest away!

Your charms in harmless childhood lay,
    Like metals in the mine.
Age from no face took more away,
    Than Youth concealed in thine!

But as your charms insensibly
    To their perfection prest;
Fond Love, as unperceived, did fly,
    And in my bosom rest.

My Passion, with your beauty grew;
    And Cupid, at my heart
Still (as his mother favoured you)
    Threw a new flaming dart!

Each gloried in their wanton part!
    To make a Lover, he
Employed the utmost of his art!
    To make a Beauty, she!

Though now I slowly bend to love,
   Uncertain of my fate;
If your fair self, my chains approve;
   I shall my freedom hate!

Lovers, like dying men, may well,
   At first, disordered be;
Since none alive can truly tell
   What fortune they must see!

———

PHILLIS! men say, That all my vows
   Are to thy fortune paid!
Alas! my heart he little knows;
   Who thinks my love a trade!

Were I, of all these woods the Lord!
   One berry from thy hand,
More solid pleasure would afford,
   Than all my large command.

My humble love has learned to live
   On what the nicest Maid,
Without a conscious blush, may give
   Beneath the myrtle shade!

Of costly food it has no need,
   And nothing will devour;
But, like the harmless bee, can feed,
   And not impair the flower.

A spotless innocence like thine,
  May such a flame allow!
Yet thy fair name for ever shine,
  As doth thy beauty now.

I heard thee wish, My lambs might stray
  Safe from the fox's power!
Though ev'ry one becomes his prey,
  I'm richer than before!

———

Love still has something of the sea;
  From whence his mother rose.
No time, his slaves from doubt can free;
  Nor give their thoughts repose.

They are becalmed in clearest days;
  And in rough weather tost!
They wither under cold delays;
  Or are in tempests lost!

One while, they seem to touch the port:
  Then straight into the Main,
Some angry wind, in cruel sport,
  Their vessel drives again!

At first, Disdain and Pride they fear!
　　Which if they chance to 'scape,
Rivals and Falsehood soon appear
　　In a more dreadful shape!

By such degrees, to Joy they come;
　　And are so long withstood,
So slowly they receive the sum,
　　It hardly does them good!

[The 1673, 1693, and 1701 texts have here]

'Tis cruel to prolong a pain!
　　And to defer a bliss,
Believe me, gentle HERMIONE!
　　No less inhuman is.

[In the 1707 text, this stanza is replaced by the following one.]

'Tis cruel to prolong a pain!
　　And to defer a joy,
Believe me, gentle CELEMENE!
　　Offends the wingèd Boy!

A hundred thousand oaths, your fears,
　　Perhaps, would not remove!
And if I gazed a thousand years,
　　I could no deeper love! . . . .

TO CELIA.

You tell me, CELIA! you approve;
Yet never must return my love!
An answer that my hope destroys;
And, in the cradle, wounds our joys!
To kill at once, what needs must die,
None would, to birds and beasts deny!
How can you then so cruel prove
As to preserve, and torture, love?
That beauty Nature kindly meant
For her own pride, and our content;
Why should the tyrant HONOUR make
[Our greatest torment? Let us break
His yoke; and that base Power disdain!
Which only keeps the good in pain.]
Our cruel, undeservèd wrack?
In Love and War, th' Impostor does
The best to greatest harms expose!

Come then, my CELIA! let 's no more
This Devil for a God adore!
Like foolish Indians, we have been;
Whose whole religion is a sin.

[If we, the Laws of LOVE had kept,
And not in Dreams of Honour slept;
He would have, surely, long ere this,
Have crowned us with the highest bliss!
Our joy had then been as complete
As now our folly has been great!]
  Let 's lose no time then; but repent!
LOVE welcomes best a penitent!

HER ANSWER.

THYRSIS! I wish, as well as you,
To Honour there were nothing due!
Then would I pay my debt of Love
In the same coin that you approve:
Which, now, you must in Friendship take!
'Tis all the payment I can make!
Friendship so high; that I must say,
'Tis rather Love with some allay.
  And rest contented! since that I,
As well myself, as you, deny!
Learn then of me, bravely to bear
The want of what you hold most dear!
And that which Honour does in me;
Let my example work in thee!

THE INDIFFERENCE.

THANKS, fair URANIA! to your scorn,
I now am free as I was born!
Of all the pain that I endured
By your late coldness, I am cured!

In losing me, proud Nymph! you lose
The humblest slave your beauty knows!
In losing you, I but throw down
A cruel tyrant from her throne!

My ranging love did never find
Such charms of person and of mind!
Y' have Beauty, Wit! and all things know
But where you should your love bestow!

I, unawares, my freedom gave,
And to those tyrants grew a slave;
Would you have kept what you had won,
You should have more compassion shown!

Love is a burden which two hearts,
When equally they bear their parts,
With pleasure carry: but no one,
Alas! can bear it long alone!

I'm not of those, who court their pain,
And make an idol of Disdain!
My hope in love does ne'er expire
But it extinguishes desire!

Nor yet of those who, ill received,
Would have it otherwise believed;
And where their love could not prevail,
Take the vain liberty to rail!

Whoe'er would make his Victor less,
Must his own weak defence confess;
And while her power he does defame;
He poorly doubles his own shame!

Even that, malice does betray,
And speak concern another way;
And all such scorn in men is but
The smokes of fires ill put out!

He's still in torment, whom the rage
To detraction does engage!
In love, Indifference is, sure,
The only sign of perfect cure!

[Yet, cruel Fair! if thou canst prove
As happy in some other Love,
As I could once have done in thine;
The sun, on happier does not shine!]

WALKING among thick shades alone,
    I heard a distant voice;
Which, sighing, said, 'Now She is gone;
    I'll make no second choice!'

I looked, and saw it was a Swain;
    Who, to the flying wind,
Did of some neighbouring Nymph complain,
    Too fair, and too unkind.

He told me, How he saw her first:
    And with what gracious eyes
And gentle speech, that flame She nurst;
    Which since She did despise.

His vows She did as fast receive
    As He could breathe them to her.
Love in her eyes proclaimed her leave,
    That He alone should woo her.

They fed their flocks still near one place,
    And at one instant met;
He, gazing on her lovely face,
    Fell deeper in the net.

She seemed of her new captive glad;
  Proud of his bondage He:
No Lover, sure, a prospect had
  Of more felicity!

But the false Maid, or never loved,
  Or gave so quickly o'er,
Ere his was to the height improved,
  Her kindness was no more.

Even her dissemblings She let fall;
  And made him plainly see
That, though his heart She did enthral,
  Her own was ever free.

Now (lest his care should pity move)
  She shuns his very sight!
And leaves him to that hopeless love
  She did create in spite.

Her name, I could not make him tell;
  Though vowing him my aid.
He said, He never would reveal
  In life, or death, the Maid!

THE COMPLAINT.

WHEN fair AURELIA first became
    The mistress of his heart,
So mild and gentle was her reign,
    THYRSIS, in hers had part.

Reserves and care he laid aside,
    And gave a loose to Love;
The headlong course he must abide,
    How steep soe'er it prove!

At first, Disdain and Pride he feared!
    But they being overthrown,
No second foe a while appeared;
    And he thought all his own.

He thought himself a happier man
    Than ever loved before!
Her favours still his hopes outran;
    Yet still he loved the more!

LOVE smiled at first; then, looking grave,
    Said, 'THYRSIS, leave to boast!
More joy than all her kindness gave,
    Her fickleness will cost!'

He spoke : and from that fatal time,
  All THYRSIS did, or said,
Appeared unwelcome, or a crime,
  To the ungrateful Maid.

[Then he, despairing of her heart,
  Would fain have had his own!
LOVE answered, 'Such a Nymph could part
  With nothing she had won!']

———

NOT, CELIA! that I juster am,
  Or better than the rest!
For I would change, each hour, like them;
  Were not my heart at rest!

But I am tied to very Thee,
  By every thought I have!
Thy face I only care to see!
  Thy heart I only crave!

All that in Woman is adored,
  In thy dear self I find!
For the whole Sex can but afford
  The Handsome and the Kind!

Why then, should I seek farther store;
  And still make love anew?
When Change itself can give no more,
  'Tis easy to be true!

———

INDIFFERENCE EXCUSED.

Love, when 'tis true, needs not the aid
   Of sighs, nor oaths, to make it known!
And to convince the cruel'st Maid,
   Lovers should use their love alone!

Into their very looks 'twill steal!
   And he that most will hide his flame,
Does, in that care, his pains reveal!
   Silence itself can love proclaim!

This, my Aurelia! made me shun
   The paths that common Lovers tread!
Whose guilty Passions are begun
   Not in their heart; but in their head!

I could not sigh; and with crossed arms
   Accuse your rigour, and my fate!
Nor tax your beauty with such charms
   As Men adore, and Women hate!

But careless lived, and without art;
   Knowing, my love you must have spied!
And thinking it a foolish part,
   To strive to shew, what none can hide!

PHILLIS is my only joy!
  Faithless as the winds, or seas,
Sometimes coming, sometimes coy;
  Yet she never fails to please!
      If with a frown
      I am cast down;
      PHILLIS, smiling
      And beguiling,
Makes me happier than before!

Though, alas! too late I find,
  Nothing can her fancy fix;
Yet the moment she is kind,
  I forgive her all her tricks!
      Which though I see,
      I can't get free!
      She deceiving,
      I believing;
What need Lovers wish for more!

———

CLORIS! I cannot say, your Eyes
Did my unwary heart surprise!
Nor will I swear, it was your Face,
Your Shape, or any nameless grace!
For you are so entirely fair,
To love a part, injustice were!

No drowning man can know which drop
Of water, his last breath did stop!
So when the stars in heaven appear,
And join to make the night look clear.
The light we no one's bounty call :
But the obliging gift of all!

He that does Lips, or Hands, adore,
Deserves them only; and no more!
But I love all and every part!
And nothing less can ease my heart!
CUPID, that Lover weakly strikes;
Who can express what 'tis he likes!

———

PHILLIS! you have enough enjoyed
   The pleasures of disdain!
Methinks, your pride should now be cloyed,
   And grow itself again.
Open to LOVE your long-shut breast,
And entertain its sweetest guest!

LOVE heals the wounds that Beauty gives;
   And can ill usage slight.
He laughs at all that Fate contrives;
   Full of his own delight.
We, in his chains, are happier far
Than Kings themselves, without them, are!

Leave then to tame Philosophy,
  The Joys of Quietness!
With me, into LOVE's Empire fly;
  And taste my happiness!
Where even tears and sighs can show
Pleasures, the cruel never know!

———

GET you gone! You will undo me!
If you love me, don't pursue me!
Let that inclination perish,
Which I dare no longer cherish!
  With harmless thoughts I did begin;
But, in the crowd, LOVE entered in.
I knew him not; he was so gay,
So innocent, and full of play!
At every hour, in every place,
I either saw, or formed, your face!
All that in Plays was finely writ;
Fancy, for you and me did fit.
My dreams at night were all of you;
Such as till then, I never knew!
  I sported thus with young Desire;
Never intending to go higher!
But now his teeth and claws are grown;
Let me that fatal lion shun! . . .

———

If Music be the food of Love;
    Sing on, till I am filled with joy!
For then my list'ning soul you move
    To pleasures that can never cloy!
Your eyes, your mien, your tongue, declare
That you are Music everywhere!

Pleasures invade both eye and ear;
    So fierce the transports are, they wound!
And all my senses feasted are,
    Though yet the treat is only sound!
    Sure, I must perish by your charms,
    Unless you save me in your arms!

TO STREPHON.

I STROVE in vain! Here, take my heart!
   But do not think your thanks are due!
For I had first tried ev'ry art,
   Th' invading Passion to subdue!
For succour, fled to Wit and Pride;
   But both, alas, their aid denied!
And Reason too, her weakness has confest;
Unable to dislodge th' imperious guest!

How swiftly does the poison spread!
   How soon 't has seized each noble part!
Wildly it rages in my head!
   Like tides of fire, consumes my heart!
Yet think not, that you conqu'ror are
   By the wise conduct of the war!
There was a traitor took your part within;
And gave you, STREPHON! what you could not win!

TO A COQUETTE BEAUTY.

FROM Wars and Plagues, come no such harms
As from a Nymph so full of charms!
So much sweetness in her face,
In her motions such a grace,
In her kind inviting eyes
Such a soft enchantment lies:
That we please ourselves too soon ;
And are with vain hopes undone!

After all her softness, we
Are but slaves ; while she is free!
Free, alas, from all desire,
Except to set the World on fire!

Thou, fair Dissembler! dost but thus
Deceive thyself, as well as us!
Like ambitious Monarchs, thou
Wouldst rather force mankind to bow,
And venture o'er the world to roam ;
Than govern with content at home.
But, trust me, CELIA! trust me, when
APOLLO's self inspires my pen!

250

One hour of Love's Delights outweighs
Whole years of Universal Praise!
And one adorer kindly used
Is of more use than crowds refused!

For what does Youth and Beauty serve?
Why, more than all your Sex deserve?
Why such soft alluring arts
To charm our eyes; and melt our hearts?
By our loss; you nothing gain!
Unless you love; you please in vain!

———

FROM all uneasy Passions free,
Revenge, Ambition, Jealousy;
Contented, I had been too blest,
If Love and You would let me rest!
Yet that dull life I now despise!
      Safe from your eyes,
I feared no griefs; but O, I found no joys!

Amidst a thousand soft desires,
Which Beauty moves, and Love inspires;
I feel such pangs of jealous fear,
No heart, so kind as mine, can bear!
Yet I'll defy the worst of harms!
      Such are those charms;
'Tis worth a life, to die within your arms!

THROUGH mournful shades, and solitary groves,
Fanned with the sighs of unsuccessful Loves,
  Wild with despair, young THYRSIS strays,
Thinks over all AMYRA's heavenly charms,
Thinks he now sees her in another's arms.
Then at some willow's root himself he lays,
      The loveliest, most unhappy, Swain:
And thus, to the wild woods he does complain.

'How art thou changed, O, THYRSIS! since the time
When thou couldst love, and hope, without a crime!
  When Nature's pride, and Earth's delight
(As through her shady evening grove she past,
And a new day did all around her cast),
Could see, nor be offended at the sight,
      The melting, sighing, wishing, Swain!
That now must never hope to wish again!

'Riches and Titles! why should they prevail
Where Duty, Love, and Adoration fail?
  Lovely AMYRA! shouldst thou prize
The empty noise that a fine Title makes,
Or the vile trash that with the vulgar takes,
Before a heart that bleeds for thee, and dies?
      Unkind! but pity the poor Swain
Your rigour kills; nor triumph o'er the slain!'

A PASTORAL.

CÆLIA AND DORINDA.

[*UNFINISHED.*]

WHEN first the young ALEXIS saw
CÆLIA to all the plain give law—
The haughty CÆLIA, in whose face
Love dwelt with Fear, and Pride with Grace;
When ev'ry Swain he saw submit
To her commanding eyes and wit:
How could th' ambitious Youth aspire
To perish by a nobler fire!
  With all the power of Verse he strove,
The lovely Shepherdess to move!
Verse, in which the Gods delight;
That makes Nymphs love, and Heroes fight!
Verse, that once ruled all the plain!
Verse, the wishes of a Swain!
How oft has THYRSIS' pipe prevailed;
Where EGON's flocks and herds have failed!
Fair AMARYLLIS! was thy mind
Ever to DAMON's wealth inclined?
Whilst LYCIDAS his gentle breast,
With love, and with a Muse possest,
Breathed forth in Verse his soft desire,
Kindling in thee his gentle fire!

⋅ ⋅ ⋅ ⋅ ⋅ ⋅ ⋅

You say, You love! Repeat again,
  Repeat th' amazing sound!
Repeat the ease of all my pain,
  The cure of ev'ry wound!

What you, to thousands have denied;
  To me you freely give!
Whilst I in humble silence died,
  Your mercy bids me live!

So upon Latmos' top, each night,
  ENDYMION sighing lay,
Gazed on the Moon's transcendent light,
  Despaired, and durst not pray!

But divine CYNTHIA saw his grief,
  Th' effect of conqu'ring charms;
Unasked, the Goddess brings relief,
  And falls into his arms!

THE MAD LOVER.

I'LL from my breast tear fond desire,
  Since LAURA is not mine;
I'll strive to cure the am'rous fire,
  And quench the flame with wine!

Perhaps, in groves and cooling shade,
  Soft slumbers I may find;
There, all the vows to LAURA made,
  Shall vanish with the wind!

The speaking strings and charming Song,
  My Passion may remove!
*O, Music will the pain prolong,
  And is the food of Love!*

I'll search Heaven, Earth, Hell, Seas, and Air;
  And that, shall set me free!
*O, Laura's image will be there,
  Where Laura will not be!*

My soul must still endure the pain;
  And with fresh torment rave!
For none can ever break the chain,
  That once was Laura's slave!

———

TO CÆLIA.

The cruel Cælia loves, and burns
  In flames she cannot hide!
Make her, dear Thyrsis! cold returns!
  Treat her with scorn and pride!

You know the captives she has made;
  The torment of her chain!
Let her, let her be once betrayed;
  Or rack her with disdain!

See, tears flow from her piercing eyes!
  She bends her knees divine!
Her tears, for Damon's sake despise!
  Let her kneel still for mine!

Pursue thy conquest, charming Youth!
  Her haughty beauty vex,
Till trembling Virgins learn this truth,
  Men can revenge their sex!

A GENTLEMAN TO HIS WIFE.

WHEN your kind wishes first I sought,
  'Twas in the dawn of Youth!
I toasted you! For you, I fought!
  But never thought of truth.

You saw, how still my fire increased;
  I grieved to be denied!
You said, Till I to wander ceased,
  You'd guard your heart with pride!

I, that once feigned too many lies,
  In height of Passion, swore
By you and other deities,
  That I would range no more!

I've sworn! and therefore now am fixed;
  No longer false and vain!
My Passion is with Honour mixed;
  And both shall ever reign!

ON HIS MISTRESS DROWNED.

SWEET stream! that dost, with equal pace,
Both thyself fly and thyself chase,
　　Forbear a while to flow;
　　And listen to my woe!

Then go, and tell the sea, That all its brine
　　　Is fresh, compared to mine!
Inform it, That the gentler Dame,
Who was the life of all my flame,
　　In th' glory of her bud,
　　Has passed the fatal flood!
Death, by this only stroke, triumphs above
　　　The greatest power of Love!

Alas! alas! I must give o'er!
My sighs will let me add no more!
Go on, sweet stream! and henceforth rest
No more than does my troubled breast!
And if my sad complaints have made thee stay;
　　These tears, these tears, shall mend thy way!

SPITE of thy Godhead, powerful LOVE!
   I will my torments hide!
For what avails, if life must prove
   A sacrifice to Pride!

PRIDE! thou'rt become my Goddess now!
   To thee, I'll altars raise!
To thee, each morn, I pay my vow;
   And offer every tear!

But O, should PHILOMEL
   Once take your injured part;
I soon should cast that idol down,
   And offer him my heart!

———

How hardly I concealed my tears!
　How oft did I complain!
When, many tedious days, my fears
　Told me, I loved in vain.

But now, my joys as wild are grown,
　And hard to be concealed!
Sorrow may make a silent moan;
　But Joy will be revealed!

I tell it to the bleating flocks!
　To every stream and tree!
And bless the hollow murmuring brooks,
　For echoing back to me!

Thus, may you see, with how much joy,
　We want! we wish! believe!
'Tis hard such Passion to destroy;
　But easy to deceive!

AN ACCOUNT OF

THE GREATEST ENGLISH POETS.

To Mr. H[ENRY] S[ACHEVERELL], April 3, 1694.

Since, dearest HARRY ! you will needs request
A Short Account of all the Muse-possest,
That, down from CHAUCER's days to DRYDEN's times,
Have spent their noble rage in British rhymes,
Without more Preface, wrote in formal length
To speak the Undertaker's want of strength,
I'll try to make their sev'ral Beauties known ;
And show their verses' worth, though not my own.

LONG had our dull forefathers slept supine,
Nor felt the raptures of the tuneful Nine ;
Till CHAUCER first, a merry Bard, arose,
And many a Story told in rhyme and prose.
But age has rusted what the Poet writ,
Worn out his language, and obscured his wit.
In vain, he jests in his unpolished strain ;
And tries to make his Readers laugh, in vain!

Old SPENSER next, warmed with poetic rage,
In antique tales amused a barbarous Age ;
An Age that, yet uncultivate and rude,
Where'er the Poet's fancy led, pursued

260

Through pathless fields and unfrequented floods,
To dens of dragons, and enchanted woods.
But now the mystic tale, that pleased of yore,
Can charm an understanding Age no more!
The long-spun Allegories fulsome grow,
While the dull Moral lies too plain below.
We view well-pleased, at distance, all the sights
Of Arms and palfreys, cattle, Fields, and fights,
And Damsels in distress, and courteous Knights.
But when we look too near, the shades decay;
And all the pleasing landscape fades away.

Great COWLEY, then, (a mighty Genius!) wrote,
O'errun with Wit, and lavish of his Thought.
His turns too closely on the Reader press!
He more had pleased us, had he pleased us less!
One glitt'ring thought no sooner strikes our eyes
With silent wonder; but new wonders rise!
As, in the Milky Way, a shining white
O'erflows the heavens, with one continued light,
That not a single star can show his rays;
While, jointly, all promote the common blaze.
Pardon, great Poet! that I dare to name
Th' unnumbered Beauties of thy Verse with blame!
Thy fault is only Wit in its excess;
But Wit like thine, in any shape will please!
What Muse but thine, could equal hints inspire;
And fit the deep-mouthed PINDAR to thy lyre!
PINDAR (whom others, in a laboured strain
And forced expression, imitate in vain!),

Well-pleased, in thee he soars with new delight,
And plays in more unbounded verse, and takes a
    nobler flight!

Blest man! whose spotless life, and charming Lays,
Employed the tuneful Prelate in thy praise.
Blest man! who now shall be for ever known
In SPRAT's successful labours, and thy own.

But MILTON next, with high and haughty stalks,
Unfettered in majestic Numbers walks.
No vulgar Hero can his Muse engage;
Nor Earth's wide scene confine his hallowed rage.
    See! See! He upward springs! and, tow'ring high,
Spurns the dull·Province of Mortality!
Shakes Heaven's eternal throne, with dire alarms;
And sets the Almighty Thunderer in Arms!
    Whate'er his pen describes, I more than see!
Whilst every verse, arrayed· in majesty,
Bold, and sublime, my whole attention draws;
And seems above the critics' nicer laws.
    How are you struck with terror and delight,
When Angel with Archangel copes in fight!
When great MESSIAH's outspread Banner shines;
How does the Chariot rattle in his lines!
What sounds of brazen wheels, what thunder, scare
And stun the Reader, with the din of war!
With fear, my spirits and my blood retire,
To see the Seraphs sunk in clouds of fire!
    But when, with eager steps, from hence I rise

262

And view the first gay scenes of Paradise;
What tongue! what words of rapture! can express
A Vision so profuse of pleasantness.
   O, had the Poet ne'er profaned his pen
To varnish o'er the guilt of faithless men;
His other Works might have deserved applause:
But, now, the Language can't support the Cause!
While the clean current, though serene and bright,
Betrays a bottom odious to the sight.

   But, now, my Muse, a softer strain rehearse!
Turn ev'ry line with art, and smooth thy Verse!
The Courtly WALLER next commands thy Lays!
Muse, tune thy Verse with art to WALLER's praise!
   While tender Airs, and lovely Dames, inspire
Soft melting thoughts, and propagate desire;
So long shall WALLER's strain, our Passion move;
And SACHARISSA's beauties kindle love!
   Thy Verse, harmonious Bard! and flatt'ring Song
Can make the vanquished great! the coward strong!
Thy Verse can show e'en CROMWELL's innocence;
And compliment the storms that bore him hence!
   O, had thy Muse not come an Age too soon;
But seen great NASSAU on the British throne!
How had his triumphs glittered in thy page;
And warmed thee to a more exalted rage!
What Scenes of Death and Horror had we viewed;
And how had Boyne's wide current reeked in blood!
   Or if MARIA's charms, thou wouldst rehearse
In smoother Numbers and a softer Verse,

Thy pen had well described her graceful air;
And GLORIANA would have seemed more fair!

Nor, must ROSCOMMON pass neglected by,
That makes e'en Rules a noble poetry!
Rules, whose deep sense and heavenly Numbers show
The best of Critics, and of Poets too!

Nor, DENHAM! must we e'er forget thy strains,
While Cooper's Hill commands the neighb'ring plains.

But see, where art-ful DRYDEN next appears,
Grown old in rhyme; but charming e'en in years!
Great DRYDEN next! whose tuneful Muse affords
The sweetest Numbers, and the fittest words.
Whether in Comic sounds, or Tragic Airs,
She forms her voice; she moves our smiles, or tears!
If Satire, or Heroic strains, she writes;
Her Hero pleases, and her Satire bites!
From her, no harsh un-art-ful Numbers fall;
She wears all dresses, and she charms in all!
How might we fear our English Poetry,
That long has flourished, should decay with thee;
Did not the Muses' other Hope appear,
Harmonious CONGREVE! and forbid our fear.
CONGREVE, whose fancy's unexhausted store
Has given already much; and promised more!
CONGREVE shall still preserve thy fame alive;
And DRYDEN's Muse shall in his friend survive!

I'm tired with rhyming, and would fain give o'er;
But Justice still demands one labour more!
   The noble MONTAGU remains unnamed;
For wit, for humour, and for judgement famed!
To DORSET, he directs his art-ful Muse
In Numbers such as DORSET's self might use!
Now, negligently graceful, he unreins
His Verse, and writes in loose familiar strains.
Now, NASSAU's Godlike acts adorn his lines;
And all the Hero in full glory shines!
We see his army set in just array;
And Boyne's dyed waves run purple to the sea.
Nor Simois, choked with men, and Arms, and blood;
Nor rapid Xanthus' celebrated flood;
Shall longer be the Poet's highest themes!
Though Gods and Heroes fought promiscuous in their
    streams.
   But now, to NASSAU's secret counsels raised;
He aids the Hero; whom before he praised!

I've done, at length! and now, dear friend, receive
The last poor present that my Muse can give!
I leave the arts of Poetry and Verse
To them, that practise them with more success!
Of greater truths, I'll now prepare to tell!
And so, at once, dear friend! and Muse! farewell!

Thomas D'Urfey.

St. George.

Genius of England! from thy pleasant bower of bliss,
  Arise, and spread thy sacred wings!
    Guard from foes the British State!
    Thou, on whose smile does wait
    Th' uncertain happy fate
    Of Monarchies and Kings.

The Genius of England.

Then, follow, brave Boys! to the wars!
  The laurel, you know, 's the prize!
Who brings home the noblest scars,
  Looks finest in Celia's eyes!
Then, shake off the slothful ease!
  Let Glory inspire your hearts!
Remember, a Soldier in war, and in peace,
  Is the noblest of all other Arts!

———

A SONG

IN PRAISE OF ARMS AND SOLDIERY.

SING, all ye Muses! your lutes strike around!
When a Soldier 's the Story, what tongue can want
    sound?
Who danger disdains, wounds, bruises, and pains;
When the honour of fighting is all that he gains!
Rich profit comes easy in cities of store;   [roar!
But the gold is earned hard, where the cannons do
Yet see, how they run, at the storming a town,
Through blood and through fire, to take the Half-Moon!
     They scale the high wall,
     Whence they see others fall;
Their hearts' precious darling, bright Glory, pursuing:
Though death 's under foot, and the mine is just blowing
It springs! Up they fly! Yet more still supply!
As Bridegrooms, to marry; they hasten, to die!
Till Fate claps her wings, and the glad tidings brings
Of the Breach being entered; and then they're all Kings!
     Then happy 's She; whose face
     Can win a Soldier's grace!
     They range about in State,
     Like Gods disposing Fate!
     No luxury in peace,
     Nor pleasure in excess,
Can parallel the joys, the martial Hero crown;
When, flushed with rage, and forced by want, he
    storms a wealthy town.

———

'DAMON! let a friend advise ye!
Follow CLORIS; though she flies ye!
Though her tongue, your suit is slighting;
Her kind eyes, you'll find inviting!
Women's rage, like shallow water,
Does but shew their hurtless nature!
When the stream seems rough and frowning,
There is still least fear of drowning!'

'Let me tell th' adventurous stranger,
In our calmness, lies our danger!
Like a river's silent running,
Stillness shews our depth and cunning!
She that rails you into trembling,
Only shews her fine dissembling!
But the fawner, to abuse ye,
Thinks you fools; and so will use ye!'

———

I FOLLOWED fame, and got renown!
I ranged all o'er the Park and Town!
I haunted Plays; and there grew wise,
Observing my own modish vice.
Friends and Wine I next did try;
Yet I found no solid joy!
Greatest pleasures seemed too small,
Till SYLVIA made amends for all!

268

But see the state of human bliss!
How vain our best contentment is!
As of my joy she was the chief;
So was she too my greatest grief!
Fate, that I might be undone,
Dooms this Angel but for one!
And, alas! too plain I see,
That I am not the happy He!

———

SOME thirty, or forty, or fifty, at least,
    Or more, I have loved in vain, in vain!
But if you'll vouchsafe to receive a poor guest;
    For once, I will venture again, again!

How long I shall be in this mind, this mind,
    Is totally in your own power!
All my days I can pass with the Kind, the Kind;
    But I'll part with the Proud in an hour!

Then, if you'll be good-natured, and civil, and civil;
    You'll find I can be so too, so too!
But if not, you may go, you may go to the Devil;
    Or the Devil may come to you, to you!

———

A DIALOGUE
BETWEEN ZEPHYRUS AND IRIS.

ZEPHYRUS. AH! what happy days and nights
        The fond Lover
        Doth discover,
When his Mistress smiles upon him!
To the heaven of sweet delights,
        Kind desire
        Mounts him higher,
Every moment she looks on him!
'Tis the noblest gift of JOVE!
'Tis the greatest joy above!
Let us, then, for ever love!
        Ever love! ever love!
Let us, then, for ever love!

IRIS.   *Bliss beyond all thought she feels;*
        Whose kind Wooer
        Does pursue her,
    With a true and constant Passion!
    Panting joy each pulse reveals!
        All her glances
        Are advances;
    When love rules her inclination!

Pray we then to mighty Jove,
That our flames may ne'er remove!
But for ever, let us love!
 Let us love! let us love!
But for ever, let us love!

———

SOLON'S SONG.

TANTIVY! tivy! tivy! tivy! high and low.
Hark! Hark! how the merry merry horn does blow
As through the lanes and the meadows we go,
 As Puss has run over the Down,   [Spring,
When Ringwood, and Rockwood, and Jowler, and
And Thunder, and Wonder, made all the woods ring;
And horsemen and footmen, hey ding a ding, ding!
Who envies the splendour and state of a Crown!

Then, follow, follow, follow, follow, jolly Boys!
Keep in with the beagles, now whilst the scent lies!
The fiery-faced God is just ready to rise;
 Whose beams all our pleasure control,
Whilst over the mountains and valleys we roll;
And Wat's fatal knell in each hollow we toll!
And, in the next cottage, top off a brown bowl!
What pleasure, like Hunting, can cherish the soul?

———

LOVE'S REVENGE.

THE world was hushed, and Nature lay
    Lulled in a soft repose,
As I, in tears, reflecting lay
    On CHLOE's faithless vows,
The God of Love, all gay, appeared
    To heal my wounded heart.
New pangs of joy my soul endeared,
    And pleasure charmed each part.
'Fond man!' said he, 'here end thy woe!
Till they, my power and justice know,
The foolish Sex will all do so!

'And, for thy ease, believe no bliss
    Is perfect without pain.
The fairest summer hurtful is,
    Without some showers of rain.
The joys of Heaven who would prize,
    If men too cheaply bought?
The dearest part of mortal joys
    Most charming is when sought:
And though with dross, true love they pay,
Those that know finest metals say,
"No gold will coin without allay!"

'But that the generous Lover may
    Not always sigh in vain,
The cruel Nymph that kills to-day,
    To-morrow shall be slain!'

The little God no sooner spoke,
  But from my sight he flew;
And I, that groaned with CHLOE's yoke,
  Found LOVE's revenge was true.
Her proud hard heart too late did turn, ⎫
With fiercer flames than mine did burn; ⎬
Whilst I as much began to scorn! ⎭

A SCOTCH SONG.

JOCKEY was a dowdy lad,
  And JEMMY swarth and tawny;
They, my heart no captive made,
  For that was prize to SAWNEY!
JOCKEY wooes, and sighs, and sues;
  And JEMMY offers money:
Weel! I see, they both love me;
  But I love only SAWNEY!

JOCKEY high his voice can raise;
  And JEMMY tunes the viol:
But when SAWNEY pipes sweet Lays,
  My heart kens no denial!
Yen he sings, and t' other strings;
  Though sweet, yet only teize me!
SAWNEY's flute can only do 't;
  And pipe a tune to please me!

THE DIRGE OF CHRYSOSTOME.

YOUNG CHRYSOSTOME had virtue, sense,
    Renown, and manly grace :
Yet all, alas ! were no defence
    Against MARCELLA's face !
His love, that long had taken root,
    In doubt's cold bed was laid ;
Where, she not warming it to shoot,
    The lovely plant decayed.

Had coy MARCELLA owned a soul
    Half beauteous as her eyes,
Her judgement had her soul controlled ;
    And taught her how to prize.
But Providence, that formed the Fair
    In such a charming skin,
Their outside made their only care ;
    And never looked within !

DIRGE.

Sleep, poor Youth ! Sleep in peace,
    Relieved from love and mortal care !
Whilst we, that pine in Life's disease,
    Uncertain blessed, less happy are !
Couched in the dark and silent grave ;
    No ills of Fate thou now canst fear !
In vain, would tyrant Power enslave ;
    Or scornful Beauty be severe !

274

Wars, that do fatal storm disperse,
  Far from thy happy mansion keep!
Earthquakes, that shake the Universe,
  Can't rock thee into sounder sleep!
With all the charms of Peace possest,
  Secure from Life's torment, or pain;
Sleep, and indulge thyself with rest!
  Nor dream thou e'er shall rise again!

CHORUS.

Past is thy fear of future doubt!
  The sun is from the Dial gone!
The sands are sunk, the Glass is out!
  The folly of the Farce is done!

ALTISIDORA'S SONG.

LOVE. FROM rosy bowers, where sleeps the God of Love,
    Hither, ye little waiting CUPIDS, fly!
Teach me, in soft melodious strains, to move
    With tender Passion, my heart's darling Joy!
Ah! let the Soul of Music tune my voice
To win dear STREPHON! who my soul enjoys.

      Or, if more influencing,
        Be doing something airy!
          With a hop, and a bound,
          And a frisk from the round,
        I'll trip, trip like a Fairy!

As when on Ida dancing
Were three Celestial Bodies;
  With an Air and a face,
  And a shape and a grace,
Let me charm like Beauty's Goddess!

MELANCHOLY. Ah! 'tis in vain! 'Tis all, 'tis all in vain!
Death and Despair must end the fatal pain!
Cold, cold Despair, disguised like snow and rain,
Falls on my breast! Black winds, in tempests
My veins all shiver, and fingers glow! [blow!
My pulse beats a Dead March for lost repose;
And to a solid lump of ice my poor fond heart
    is froze!

PASSION. O, say, ye Powers! My peace to crown,
Shall I thaw myself; and drown
Amongst the foaming billows?
  Increasing all with tears I shed;
On beds of ooze, and crystal pillows,
  Lay down my love-sick head?

FRENZY. No! No! I'll straight run mad!
  That soon my heart will warm!
When once the Sense is fled;
  Love has no power to charm!
Wild, through the woods I'll fly;
  And dare some savage boar!
A thousand deaths I'll die,
  Ere thus, in vain, adore!

THE SONG OF THE PRIESTESSES,
AT THE TOMB OF ARGACES.

Sleep, ye great Manes of the dead!
Whilst our solemn Round we tread;
Whilst at our Cell, as at a shrine,
We nightly wait with rites divine!
Whilst, to adorn the tomb, we bring
The earliest glories of the Spring;
And sweetest softest Anthems sing!
The floor, with hallowed drops bedewing;
And all around fresh roses strewing.

Ye Guardian Powers, that here resort,
For ever make this Cell your Court!
If devoutest prayers invite ye,
Or Sabæan gums delight ye;
Then make this sacred Urn your care! ⎫
And nightly to this Cell repair, ⎬
To feast on frankincense and prayer! ⎭
Around we go, the floor bedewing;
Violets, pinks, and roses strewing.

———

A DIALOGUE
BETWEEN ALEXIS AND LAURA.

LAURA. ALEXIS!

ALEXIS.   Dearest!

LAURA.   *Take a kiss!*

ALEXIS. What means this unexpected bliss?
A bliss which I so oft, in vain,
Have craved; and now, unasked, obtain?

LAURA. *When to my Swain reserved I seemed;*
I loved him! kissed him; less esteemed!

ALEXIS. Dear Nymph! your female arts forbear,
With one already in the snare!
'Tis, LAURA! an unjust design,
To treat so plain a soul as mine
With Oracles! Such mystic sense
Religion fitly may dispense!
But these dark riddles mar Love's joy,
As clouds, gems in their worth destroy!

LAURA. *Then take it, on your peril, Swain!*
Since you compel me to be plain;
The kiss I gave you was in lieu
Of all love debts from LAURA due!

ALEXIS. What crimes can I have wrought, to force
This sudden and severe divorce?

LAURA. Recall, false Shepherd! what, to-day,
I heard you to DORINDA *say!*
You said, 'She did noon's light outshine!
That Beauty's Queen was less divine!'
You vowed respect to her commands;
And, Heaven forgive you! kissed her hands!

ALEXIS. You wrong me, Nymph! By Pan, you do!
That courtship was respect to you!
DORINDA's beauties well are known
To bear such likeness to your own,
That, when I made my late address,
'Twas, in that gentle Shepherdess,
The sweetness of those charms to taste,
Which so divinely LAURA graced.

LAURA. Weak Nymphs, with Men contend in vain;
Who thus their errors can maintain!

CHORUS. Wise Nature's care is here expressed,
That neither Sex should be oppressed!
Who, when to Nymphs she did commit
Commanding charms, gave Shepherds wit,
With arts and cunning, to allay
And temper Beauty's powerful sway.

THE HURRICANE.

'WHAT cheer? my Mates! Luff ho! We toil in vain!
That northern mist forebodes a hurricane!
    See, how th' expecting ocean raves!
    The billows roar before the fray!
    Untimely night devours the day!
    I' th' dead eclipse, we nought descry
But lightning's wild caprices in the sky;
And scaly monsters sparkling through the waves!
    Ply, each a hand; and furl your sails!
    Port! Hard a port! The tackle fails!
    Sound ho!' '*Five fathom, and the most!*'
'A dangerous shelf! Sh' 'as struck; and we are lost!
Speak, in the hold!' '*She leaks amain!*' 'Give o'er!
    The crazy boat can work no more!
She draws apace; and we approach no shore!
A ring, my Mates! Let 's join a ring, and so
    Beneath the deep embracing go!
Now to new worlds, we steer; and quickly shall
    arrive!
Our spirits shall mount as fast as our dull corpses
    dive!'

———

OF VICE AND VIRTUE.

Let Vice no more, in her full Train take pride!
Who follow Virtue choose a suff'ring Side!
She 's exiled now: and 'tis not strange to see
Mean souls desert afflicted Majesty!
But when just Heaven (and, sure, that time draws on!)
Restores this Empress to her starry throne;
With Crowns she will enrich her loyal few,
While Shame and Vengeance crush the rebel crew!

THE CHOICE.

Grant me, indulgent Heaven! a rural seat,
  Rather contemptible than great!
Where, though I taste Life's sweets, still I may be
  Athirst for Immortality!
I would have business; but exempt from strife!
  A private, but an active, life!
A Conscience bold, and punctual to his charge!
  My stock of Health; or Patience large!
Some books I'd have, and some acquaintance too;
  But very good, and very few!
Then (if one mortal, two such grants may crave!)
  From silent life, I'd steal into my grave!

Nahum Tate, P.L.

THE AMUSEMENT.

STREPHON. WHY weeps my SYLVIA? Prithee, why?

SYLVIA. *To think, my STREPHON once must die!*
To think withal, poor SYLVIA may,
When he 's removed, be doomed to stay!

STREPHON. Nymph! you're too lavish of your tears,
To waste them on fantastic fears!

SYLVIA. *No! for when I this life resign,*
If Fate prolong the date of thine,
The tears you'll give my funeral
Will pay me Interest, Stock, and all!

STREPHON. Not so! For should this setting light
Ne'er rise again in SYLVIA's sight;
Without a tear in mine, I'd view
Her dying eyes!

SYLVIA.   *'Tis false!*

STREPHON.   'Tis true!

SYLVIA. *Not weep! false Shepherd! Swear!*

STREPHON.   I swear,
I would not give thy hearse a tear!

SYLVIA. *Break, swelling heart! Perfidious man!*
Can you be serious? Swear again!
Yes! swear by CERES *and by* PAN!

STREPHON. Let then great PAN and CERES hear;
And punish, if I falsely swear!

SYLVIA. *Gods! Can ye hear this, and forgive?*
You may! for I have heard, and live!

STREPHON. Rage not, rash Nymph! For I have
decreed,
When SYLVIA dies—

SYLVIA.   *Speak! what?*

STREPHON.   To bleed!
I'll drain the life-blood from my heart;
But no cheap tear shall dare to start!

SYLVIA. *Kind Shepherd! Could you life despise,*
And bleed at SYLVIA's *obsequies?*

STREPHON. To CERES I appeal! for she
Knows, this has long been my decree!

SYLVIA. *Since then you could your vow fulfil,*
Swear! Swear once more, you never will!

TO MADAM S——, AT THE COURT.

Come, prithee, leave the Courts;
   And range the fields with me!
A thousand pretty rural sports,
   I'll here invent for thee!

Involved in blissful innocence,
   We'll spend the shining day,
Untouched with that mean influence
   The duller World obey!

About the flow'ry plains we'll rove,
   As gay and unconfined
As are, inspired by thee and love,
   The sallies of my mind!

Now seated by a lovely stream,
   Where beauteous Mermaids haunt,
My Song, while William is my theme,
   Shall them and thee inchant!

Then in some gentle soft retreat,
   Secure as VENUS' groves,
We'll all the charming things repeat
   That introduced our loves.

I'll pluck fresh garlands for thy brows,
   Sweet as a Zephyr's breath,
As fair and well designed as those
   The Elysium Lovers wreathe.

And like those happy Lovers, we,
   As careless and as blessed,
Shall, in each other's converse, be
   Of the whole World possessed.

Then, prithee, PHILLIS! leave the Courts;
   And range the fields with me!
Since I so many harmless sports
   Can here procure for thee!

He 's gone the bright way that his honour directs
    him !
O, all ye kind Powers ! let me beg you protect him !
He 's gone, my dear ——! and left me here mourning :
But, hang these dull thoughts ! I'll fancy him re-
    turning !
Returning, I'll think the great hero victorious,
With joy to my arms, as faithful, as glorious.
Against his bright eyes, I am sure, there 's no standing.
He looks like a God ; and moves as commanding !
With a face so angelic, the Foe will be charmed ;
The conquest were his ; though he met them disarmed !
They could not be, sure ! of a rational nature,
That would not relent at so moving a feature.
'Venus disguised,' he'll be thought, by his beauty ;
And spared from the sense of a generous duty.
    Yet when I reflect on the wounded and dying,
In spite of my courage, it sets me a sighing !
But the resolute brave, no danger can stay him !
Though I used all my charms and arts to delay him !
    Yet O, ye kind Powers ! you are bound to protect
    him ;
Since he 's gone the bright way that Glory directs him !

A BEAUTEOUS face, fine shape, engaging Air,
With all the graces that adorn the Fair;
If these could fail their so-accustomed Parts,
And not secure the conquest of our hearts;
SYLVIA has yet a vast reserve in store!
At sight, we love; but hearing, must adore!

There falls continual music from her tongue!
The wit of SAPPHO, with her art-full Song!
From Sirens, thus we lose the power to fly;
We listen to the Charm, and stay to die!
Ah! lovely Nymph! I yield! I am undone!
Your Voice has finished what your Eyes begun!

———

How! 'Mortal hate!' For what offence?
For too much love? or negligence?
The first, Who is it that denies
The fault of your victorious eyes?
As 'tis of your severer Arms
I pay no more, my tribute to your charms!

Yet I, in silence, still admire!
Have gazed, till I have stole a fire!
A mighty crime, in one you hate;
Yet who can see and shun the Fate!
Ah! let it then not mortal prove!
Not, but I'd die, to shew how much I love!

John Oldmixon.

TO CORINNA.

FAIR CORINNA! Tell me why
You are often heard to sigh?
Why your eyes are often seen
Kind as Lovers' should have been?
Tell me, Madam, what you mean?
Something does your soul employ,
Love or Anger, Grief or Joy!
By the symptoms, we discover
Something even of a Lover!

Love, like murder, will appear;
Though you take the greatest care!
Every motion will reveal
What you struggle to conceal!
Hide it not! For I perceive
When your breasts begin to heave!
When they rise, and when they fall;
Then I see, and know, it all!

They, in spite of all your art,
Tell the conflicts of your heart!
Every throb and pant repeat,
Equal time and motion beat;
But for whom your wishes grow,
That, O, that, I cannot know!

THOSE arts which common Beauties move,
   CORINNA, you despise!
You think there 's nothing wise in love,
   Or eloquent in sighs.
You laugh at ogle, cant, and Song;
   And promises abuse:
But say! (for I have courted long!)
   What methods shall I use?

We must not praise your charms and wit;
   Nor talk of dart and flame!
But sometimes you can think it fit
   To smile at what you blame.
Your Sex's forms, which you disown,
   Alas! you can't forbear!
But, in a minute, smile and frown;
   Are tender and severe!

CORINNA! let us now be free!
   No more your arts pursue;
Unless you suffer me to be
   As whimsical as you!
At last, the vain dispute desist!
   To Love resign the Field!
'Twas custom forced you to resist;
   And custom bids you yield!

TO CORINNA.

Say, Corinna! Do you find
Nothing in your bosom kind?
Is it never less severe;
Or d'ye never wish it were?
Yes! I read it in your eyes!
Hear it, know it, by your sighs!
Sighs, that gently steal their way,
Tell me all that you should say!
Tell me, when you seem serene,
You're not always calm within:
But are vexed with tumults there,
Such as oft disturb the Fair!
Say, Corinna! Is this true?
Say, for I'm a Lover too;
And can tell you, what to do!

He that 's worthy to be blessed,
Should be, first, of truth possessed!
Young and constant he must be;
Fixed like you, and fond like me!
One that all affronts can bear,
Exiles, jealousies, despair!
One on whom you may depend
For a Lover and a Friend!

Plead not now for an excuse!
Man does naught like this produce.
Justice, Madam! bids you see ⎱
All these qualities in me! ⎬
Justice tells you, I am He! ⎰

———

TO CLOE.

PRITHEE, CLOE! not so fast!
Let 's not run and wed in haste!
We've a thousand things to do!
You must fly; and I pursue!
You must frown; and I must sigh!
I intreat; and you deny!
Stay!—If I am never crost,
Half the pleasure will be lost!
Be, or seem to be, severe!
Give me reason to despair!
Fondness will my wishes cloy,
Make me careless of the joy!
Lovers may, of course, complain
Of their trouble and their pain;
But if pain and trouble cease,
Love, without it, will not please!

As the inamoured THIRSIS lay
    With his SILVIA reconciled;
Whose eyes did brighter beams display,
    While the lovely Charmer smiled;
With joy transported cried, 'My Dear!
    Let us, let us often jar!
Peace always sweetest does appear
    After sharp fatigues of war.'

'No!' said the Nymph, 'mistaken Swain!
    'Tis best, our quarrels to give o'er!
Kingdoms may jar, and close again;
    But broken Love cements no more!'

THE CHOICE.

SILVIA! of all your amorous Train,
    The Black, the Brown, or Fair,
The wealthy Lord, or humble Swain;
    For whom will you declare?

If Wealth, or Beauty do prevail;
    My claim I then resign!
If Truth, and Love, I cannot fail!
    And SILVIA must be mine!

FAIREST of thy Sex, and best,
  Admit my humble tale!
'Twill ease the torment of my breast;
  Though I shall ne'er prevail!

No fond ambition me does move
  Your favour to implore;
I ask not for return of love,
  But freedom to adore!

TO CUPID.

I KNOW thy malice, trifling Boy!
Thou wouldst my happiness destroy;
Because SEPTIMIUS wounded lies,
Not by thy darts, but ACME's eyes!
  Shake not at me, thy threat'ning dart!
But wound the cruel ACME's heart!
But, O, I fear thy deity will prove
Too weak, to thaw that icy Maid to Love!

I LATELY vowed, but 'twas in haste,
  That I no more would court
The joys which seem, when they are past,
  As dull as they are short!
I oft, to hate my Mistress swear;
  But soon my weakness find.
I make my oaths, when she 's severe;
  And break them when she 's kind!

WHEN Maids live to thirty, yet never repented ;
When all Europe's at peace, and all England contented;
When no Gamester will swear, and no bribery thrives ;
Young wives love old husbands, young husbands, old
    wives ;
When Landlords love taxes, and Soldiers love peace ;
And Lawyers forget a rich client to fleece ;
    When an old face shall please as well as a new ;
    Wives, Husbands, and Lovers will ever be true !

    When Bullies leave huffing, and Cowards, their trem-
        bling ;
And Courtiers, and Women, and Priests, their dis-
        sembling ;
When these shall do nothing against what they teach,
Pluralities hate ; and we mind what they preach ;
When Vintners leave brewing, to draw the wine pure ;
And Quacks, by their medicines, kill less than they cure ;
    When an old face shall please as well as a new ;
    Wives, Husbands, and Lovers will ever be true !

Ah! cruel Beauty! could you prove
    More tender, or less fair;
You neither would provoke my love,
    Nor cause me to despair.
But your dissembling charming eyes
    My easy hopes beguile;
And though a rock beneath them lies,
    The tempting surface smiles.

To what your Sex, on ours impose,
    My humble love complied;
And when my secret I disclosed,
    Thought modesty denied!
'Yes! sure,' said I, 'her yielding heart
    Partakes of my desire!
But nicer Honour feigns this part,
    To hide the rising fire!'

Against your mind, my suit I told,
    And slighted vows renewed;
Yet you insensibly were cold,
    And I but vainly wooed.
Then, for return, a scorn prepare;
    Or lay that frown aside!
Affected coyness I can bear;
    But hate insulting pride!

WHY this talking still of dying?
  Why this dismal look and groan?
Leave, fond Lover! leave your sighing!
  Let these fruitless arts alone!
Love 's the child of Joy and Pleasure,
  Born of Beauty, nursed with Wit.
Much amiss you take your measure,
  This dull whining way to hit. . . .

——

BY A PERSON OF QUALITY.

WHO can resist my CELIA's charms?
Her beauty wounds; and wit disarms!
When these, their mighty forces join;
What heart 's so strong but must resign?
Love seems to promise, in her eyes,
A kind and lasting Age of joys.
  But, have a care! their treason shun!
I looked, believed, and was undone!
In vain, a thousand ways I strive
To keep my fainting hopes alive!
My love can never find reward;
Since Pride and Honour are her guard.

——

TOBACCO is but an Indian weed,
Grows green in the morn; cut down at eve.
    It shows our decay.
    We are but clay!
Think of this, and take Tobacco!

The Pipe, that is so lily-white,
Wherein so many take delight,
    Is broke with a touch.
    Man's life is such!
Think of this, and take Tobacco!

The Pipe, that is so foul within,
Shews how Man's soul is stained with sin.
    It does require
    To be purged with fire!
Think of this, and take Tobacco!

The Ashes, that are left behind,
Do serve to put us all in mind,
    That unto dust,
    Return we must!
Think of this, and take Tobacco!

The Smoke, that does so high ascend,
Shews you Man's life must have an end.
    The vapour 's gone!
    Man's life is done!
Think of this, and take Tobacco!

'I SMILE at LOVE, and all his arts!'
   The charming CYNTHIA cried.
'Take heed! for LOVE has piercing darts!'
   A wounded Swain replied.
'Once free and blest, as you are now;
   I trifled with his Charms!
I pointed at his little bow;
   And sported with his Arms!
Till, urged too far, "Revenge!" he cries,
   A fatal shaft he drew.
It took its passage through your eyes;
   And to my heart it flew.

'To tear it thence, I tried in vain!
   To strive; I quickly found
Was only to increase the pain,
   And to enlarge the wound!
Ah! much too well, I fear you know
   What pain I'm to endure!
Since what your Eyes alone could do;
   Your Heart alone can cure!
And that (Grant, Heaven! I may mistake!)
   I doubt is doomed to bear
A burden, for another's sake;
   Who ill rewards its care.'

RETIREMENT.

O, WHITHER is my Love withdrawn? and where
Shall I direct my steps, to find him near;
And fear no prying eyes, nor list'ning ear?

Ah! whither shall I now repair, to find
A solitary place, to ease my mind
To him, that can my foes in fetters bind?

Ah! LORD! withhold not those resplendent beams;
Nor stop the current of those crystal streams,
Which would both heal my wounds, and cleanse my
    stains!

Spring, spring, O, Fountain of Felicity!
And let sweet Shiloh's stream flow forth! that I
May drink, and be refreshed; or else I die!

Appear, Thou great High Priest of Israel!
Who, in that vast eternity doth dwell;
Whose perfect beauty, tongue can never tell!

O, can my soul but much affected be
With those enam'ring rays that dart from thee,
Great Spring of Light! O, teach me fervently

To wait for thee! whose love is still inclined
To favour Isr'el's Seed, whose heart and mind
Is unto thee unfeignedly resigned.

Though mortals cannot see thy face and live;
Yet to thy Royal Seed Thou'rt pleased to give
This princely favour! this prerogative!

Lord! cast thine eye upon a worm; whose state
Is even like a widow desolate!
Remove, consume, all that would separate,

Or interpose betwixt my Love and me!
O, rend the veil throughout, that I may see
My soul's chief darling! Ah! Thou, Thou art he

That stand'st behind the wall; and there dost wait
To show thyself, when thine do supplicate
To see thee near! Then, dost Thou satiate

The pure in heart! although Thou please to try
Their zeal, their fervour, their sincerity
To thee; whose presence, Lord! is always nigh!

THE END OF THE DRYDEN ANTHOLOGY.

FIRST LINES AND NOTES.

Many of these Poems became immediately popular; and appeared in other contemporary editions than those here quoted, often with great variations in the texts.
All the Works herein quoted, were published in London; unless otherwise stated.
Where a text is found associated with music, (M.) is put after its date.

First Lines and Notes.

First Lines and Notes.

GLOSSARY AND INDEX.

308

Glossary and Index.

Glossary and Index.

Glossary and Index.